Agricultural Automation: Principles, Systems and Applications

Agricultural Automation: Principles, Systems and Applications

Karl Pegg

Larsen & Keller
www.larsen-keller.com

Agricultural Automation: Principles, Systems and Applications
Karl Pegg
ISBN: 978-1-64172-597-2 (Hardback)

Larsen & Keller

Published by Larsen and Keller Education,
5 Penn Plaza,
19th Floor,
New York, NY 10001, USA

Cataloging-in-Publication Data

Agricultural automation : principles, systems and applications / Karl Pegg.
     p. cm.
Includes bibliographical references and index.
ISBN 978-1-64172-597-2
1. Agriculture--Automation. 2. Farm mechanization. 3. Agricultural engineering. I. Pegg, Karl.
S675 .A37 2022
631.3--dc23

For more information regarding Larsen and Keller Education and its products, please visit the publisher's website www.larsen-keller.com

Table of Contents

Preface

The machinery which is used in agriculture and farming is known as agricultural machinery. There has been a gradual shift in the recent years to introduce technology in order to automate these machines and minimize human intervention. A few examples of the different tasks, which are being automated are monitoring moisture in the soil, seeding and harvesting. There are various software and hardware components involved in these processes. The software components are aimed at managing soil, seeds, irrigation and fertilizers to make predictions and provide information to maximize yield. The hardware components are sensors, drones, harvesting robots and seeders. The topics included in this book on agricultural automation are of utmost significance and bound to provide incredible insights to readers. It elucidates new techniques and their applications in a multidisciplinary approach. This book is an essential guide for both academicians and those who wish to pursue this discipline further.

A detailed account of the significant topics covered in this book is provided below:

Chapter 1- Agricultural automation is concerned with increasing the productivity of agricultural machinery by improving its efficiency, reliability and reducing the human intervention. This chapter has been carefully written to provide an easy understanding of agricultural automation in modern farming.

Chapter 2- Precision Agriculture is a type of farming management which is based upon observation and analysis of inter and intra-field variability in crops. It uses different tools and technologies such as GPS, sensors, variable rate seeding, weather modeling, etc. The topics elaborated in this chapter will help in gaining a better perspective about these tools and related aspects of precision agriculture.

Chapter 3- Irrigation refers to the usage of controlled amounts of water to plants and crops at regular intervals. Automatic irrigation is the system used for the operation and management of irrigation structures. The diverse applications of automatic irrigation system have been thoroughly discussed in this chapter.

Chapter 4- Some of the equipment that are used in harvest automation are grain cart, conveyor belt, haulm toppers, etc. Dehulling system and sorting system fall under the domain of post-harvest automation. The topics elaborated in this chapter will help in gaining a better perspective about harvest and post-harvest automation.

Chapter 5- Agricultural robots are used to increase the production yield for farmers. Driverless tractor, weeding robots, fruit picking robots, harvest automation robot, agricultural drones, etc. are some of these robots. This chapter discusses about these agricultural robots in detail.

Chapter 6- Automation and IoT are used in different agricultural purposes such as weed control, cloud seeding, planting seeds, harvesting, environmental monitoring and soil analysis. This chapter closely examines these applications of automation and IoT in agriculture to provide an extensive understanding of the subject.

It gives me an immense pleasure to thank our entire team for their efforts. Finally in the end, I would like to thank my family and colleagues who have been a great source of inspiration and support.

Karl Pegg

Agricultural Automation: An Introduction

Agricultural automation is concerned with increasing the productivity of agricultural machinery by improving its efficiency, reliability and reducing the human intervention. This chapter has been carefully written to provide an easy understanding of agricultural automation in modern farming.

Automation in Agriculture

Agriculture is the oldest and most important economic activity of humanity, because it provides the food, fiber and the fuel necessary for our survival. With the world population expected to reach 9 billion by 2050, agricultural production must double to meet the growing demand for food and bioenergy. Taking into account the limited resources of land, water and manpower, it is estimated that the effciency of agricultural productivity should increase by 25% to achieve this goal by limiting the growing pressure exerted by agriculture on the environment. With population growth and rising food demand, the agricultural industry knows how important it is to move quickly to ensure that supply meets demand. The agricultural industry uses robots for a wide range of tasks to meet these needs. Automation is an illumination of the human resources of the labor camp. From last two decades, the concept of industrial automation in agriculture has not developed much.

Automation to agriculture helps create the various advances in the industry and helps farmers to save time and money. Some of the scientific contributions towards the mobile robot, the flying robot and the forest robot are exclusively for agriculture. Even in developing countries, farmers are interested in using robots to take care of the fields, pick the fruit or even maintain the animal, agricultural robots must have human interaction to compensate for the problems of programming complexity. The technological challenges will soon be largely resolved and the industry will be in the process of creating and testing a corporate case, both as a team and as a service.

New developed robots will be able to monitor crops and animals. This would include the control of growth and disease models. Using Wi-Fi technology, the machine can navigate crops without damaging them, even on rough terrain. The device is designed to collect information on soil composition, temperature, humidity and the general conditions of the plants to be sent to the farmer. The mechanical design consists of the end effector, the manipulator and the clamp, the manipulator design includes the task, the economic efficiency and the required movements, the final effectors in the agricultural robot is the device located at the end of the robotic arm and use for various agricultural operations.

Robots and drones have already quietly begun to transform many aspects of agriculture. Thousands of robotic milking parlors have already been installed around the world, creating a $ 1.9 billion sector that is estimated to increase to $ 8 billion by 2023. Mobile robots are also penetrating dairy farms, helping to automate tasks such as pushing or cleaning manure. Also the automatic driving and tractor driving technologies are generalized thanks to the improvements and the cost reduction of the RTK GPS technology. In fact, more than 300k of tractors with automatic transmission or tractor was sold in 2016, reaching over 660k units a year in 2027.

The use of robotics can replace many of the ordinary and often arduous work in various sectors of farming and animal husbandry. Robot platforms can carry workers to pick fruit, and some automation can pick the fruit itself. There are also autonomous robots that can harvest crops in field. They can automate crop spraying and spread fertilizer through driverless tractor. They can even sheer sheep and milk cows. However, these farming tasks are not the only way things are automated in the agricultural field. There are also industrial applications that make life easier for those involved in agriculture.

Agricultural robotics is also quickly developing on the earth. Vision-enabled robotics implements have been in viable use for some years in organic farming. These implements follow the crop rows, identify the weeds, and aid with mechanical hoeing. The next generation of these innovative robotic implements is also in its early phase of commercial deployment.

To make a selective collection effectively, two criteria are required; the ability to detect the quality factor before harvest and the ability to collect the product of interest without damaging the rest of the crop. Most agricultural equipment is becoming larger and therefore is not suitable for this approach. Smaller and more versatile selective collection equipment is needed. Either the crop can be inspected before harvesting so that the necessary information about where the crop of interest is located, or that the harvester can have mounted sensors that can determine crop conditions. The selective harvester can harvest that ready crop, allowing the rest to mature, dry or ripen, etc. Alternatively, small harvesters of stand-alone crops could be used to selectively harvest the entire crop from a selected area and transport it to a stationary processing system that could clean, cut and perhaps pack the product. This is not a new idea, but the updating of a system that used stationary threshers from many years ago. Alternatively, a separator header could be used to collect the heads of the cereals and send them to threshing.

Agricultural automation is a continuous development. The existing research technologies give rise to the possibility of developing a completely new mechanization system to support the cultivation system based on small intelligent machines. This system replaces the complete energy on the application with intelligently targeted inputs thus reducing the cost of inputs while increasing the level of care. This can improve the economics of crop production and harvest less environmental impact.

Smart Farming — Automated and Connected Agriculture

Smart farming and precision agriculture involve the integration of advanced technologies into existing farming practices in order to increase production efficiency and the quality of agricultural products. As an added benefit, they also improve the quality of life for farm workers by reducing heavy labor and tedious tasks.

Just about every aspect of farming can benefit from technological advancements—from planting and watering to crop health and harvesting. Most of the current and impending agricultural technologies fall into three categories that are expected to become the pillars of the smart farm: Autonomous robots, drones or UAVs, and sensors and the Internet of Things (IoT).

How are these technologies already changing agriculture, and what new changes will they bring in the future?

Autonomous and Robotic Labour

Replacing human labor with automation is a growing trend across multiple industries, and agriculture is no exception. Most aspects of farming are exceptionally labor-intensive, with much of that labor comprised of repetitive and standardized tasks—an ideal niche for robotics and automation.

We're already seeing agricultural robots—or AgBots—beginning to appear on farms and performing tasks ranging from planting and watering, to harvesting and sorting. Eventually, this new wave of smart equipment will make it possible to produce more and higher quality food with less manpower.

Driverless Tractors

The tractor is the heart of a farm, used for many different tasks depending on the type of farm and the configuration of its ancillary equipment. As autonomous driving technologies advance, tractors are expected to become some of the earliest machines to be converted.

In the early stages, human effort will still be required to set up field and boundary maps, program the best field paths using path planning software, and decide other operating conditions. Humans will also still be required for regular repair and maintenance.

Nevertheless, autonomous tractors will become more capable and self-sufficient over time, especially with the inclusion of additional cameras and machine vision systems, GPS for navigation, IoT connectivity to enable remote monitoring and operation and radar and LiDAR for object detection and avoidance. All of these technological advancements will significantly diminish the need for humans to actively control these machines.

According to CNH Industrial, a company that specializes in farm equipment and previewed a concept autonomous tractor in 2016, "In the future, these concept tractors will be able to use 'big data' such as real-time weather satellite information to automatically make the best use of ideal conditions, independent of human input, and regardless of the time of day."

Seeding and Planting

Sowing seeds was once a laborious manual process. Modern agriculture improved on that with seeding machines, which can cover more ground much faster than a human. However, these often use a scatter method that can be inaccurate and wasteful when seeds fall outside of the optimal location. Effective seeding requires control over two variables: planting seeds at the correct depth, and spacing plants at the appropriate distance apart to allow for optimal growth.

Precision seeding equipment is designed to maximize these variables every time. Combining geomapping and sensor data detailing soil quality, density, moisture and nutrient levels takes a lot of the guesswork out of the seeding process. Seeds have the best chance to sprout and grow and the overall crop will have a greater harvest.

As farming moves into the future, existing precision seeders will come together with autonomous tractors and IoT-enabled systems that feed information back to the farmer. An entire field could be planted this way, with only a single human monitoring the process over a video feed or digital control dashboard on a computer or tablet, while multiple machines roll across the field.

Automatic Watering and Irrigation

Subsurface Drip Irrigation (SDI) is already a prevalent irrigation method that allows farmers to control when and how much water their crops receive. By pairing these SDI systems with increasingly sophisticated IoT-enabled sensors to continuously monitor moisture levels and plant health, farmers will be able to intervene only when necessary, otherwise allowing the system to operate autonomously.

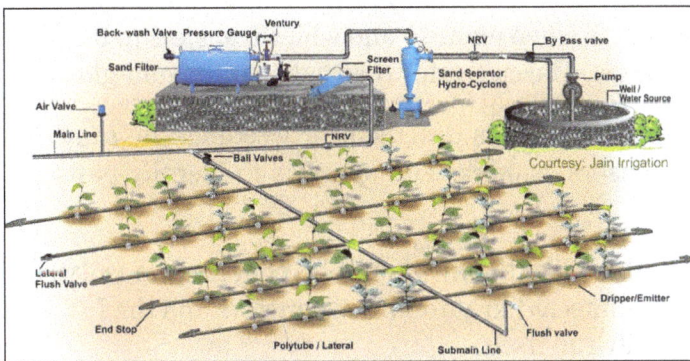

Example of an SDI system for agriculture. While current systems often require the farmer to manually check lines and monitor the pumps, filters and gauges, future farms can connect all this equipment to sensors that stream monitoring data directly to a computer or smartphone.

While SDI systems are not exactly robotic, they could operate completely autonomously in a smart farm context, relying on data from sensors deployed around the fields to perform irrigation as needed.

Weeding and Crop Maintenance

Weeding and pest control are both critical aspects of plant maintenance and tasks that are perfect for autonomous robots. A few prototypes are already being developed, including Bonirob from Deepfield Robotics, and an automated cultivator that is part of the UC Davis Smart Farm research initiative.

The Bonirob robot is about the size of a car and can navigate autonomously through a field of crops using video, LiDAR and satellite GPS. Its developers are using machine learning to teach the Bonirob to identify weeds before removing them. With advanced machine learning, or even artificial intelligence (AI) being integrated in the future, machines such as this could entirely replace the need for humans to manually weed or monitor crops.

The Bonirob farming robot.

The UC Davis prototype operates a bit differently. Their cultivator is towed behind a tractor and is equipped with imaging systems that can identify a fluorescent dye that the seeds are coated with when planted, and which transfers to the young plants as they sprout and start to grow. The cultivator then cuts out the non-glowing weeds.

While these examples are robots designed for weeding, the same base machine can be equipped with sensors, cameras and sprayers to identify pests and application of insecticides.

These robots, and others like them, will not be operating in isolation on farms of the future. They will be connected to autonomous tractors and the IoT, enabling the whole operation to practically run itself.

Harvesting from Field, Tree and Vine

Harvesting depends on knowing when the crops are ready, working around the weather and completing the harvest in the limited window of time available. There are a wide variety of machines currently in use for crop harvesting, many of which would be suitable for automation in the future.

Traditional combine, forage, and specialty harvesters could immediately benefit from autonomous tractor technology to traverse the fields. Add in more sophisticated tech with sensors and IoT connectivity, and the machines could automatically begin the harvest as soon as conditions are ideal, freeing the farmer for other tasks.

Developing technology capable of delicate harvest work, like picking fruit from trees or vegetables such as tomatoes, is where high-tech farms will really shine. Engineers are working to create the right robotic components for these sophisticated tasks, such as Panasonic's tomato-picking robot which incorporates sophisticated cameras and algorithms to identify a tomato's color, shape and location to determine its ripeness.

Another prototype for fruit picking is the vacuum-powered apple picking robot by Abundant Robotics, which uses computer vision to locate apples on the tree and determine if they are ready to harvest.

These are only a few of the dozens of up-and-coming robotic designs that will soon take over harvesting labor. Once again, with the backbone of a robust IoT system, these agbots could continuously patrol fields, check on plants with their sensors and harvest ripe crops as appropriate.

Reducing Labor, Increasing Yield and Efficiency

The core concept of incorporating autonomous robotics into agriculture remains the goal of reducing reliance on manual labour, while increasing efficiency, product yield and quality.

Unlike their forebears, whose time was mostly taken up by heavy labor, the farmers of the future will spend their time performing tasks such as repairing machinery, debugging robot coding, analyzing data and planning farm operations.

As noted with all of these agbots, having a robust backbone of sensors and IoT built into the farm's infrastructure is essential. The key to a truly "smart" farm relies on the ability of all the machines and sensors being able to communicate with each other and with the farmer, even as they operate autonomously.

Drones for Imaging, Planting and More

What farmer wouldn't want a bird's eye view of their fields? Where once this required hiring a helicopter or small aircraft pilot to fly over a property taking aerial photographs, drones equipped with cameras can now produce the same images at a fraction of the cost.

In addition, advances in imaging technologies mean that you're no longer limited to visible light and still photography. Camera systems are available spanning everything from standard photographic imaging, to infrared, ultraviolet and even hyperspectral imaging. Many of these cameras can also record video. Image resolution across all these imaging methods has increased, as well, and the value of "high" in "high resolution" continues to rise.

All these different imaging types enable farmers to collect more detailed data than ever before, enhancing their capabilities for monitoring crop health, assessing soil quality and planning planting locations to optimize resources and land use. Being able to regularly perform these field surveys improves planning for seed planting patterns, irrigation and location mapping in both 2D and 3D. With all this data, farmers can optimize every aspect of their land and crop management.

But it is not just cameras and imaging capabilities making a drone-assisted impact in the agricultural sphere—drones are also seeing use in planting and spraying.

Planting from the Air

Prototype drones are being built and tested for use in seeding and planting to replace the need for manual labor. For example, several companies and researchers are working on drones that can use compressed air to fire capsules containing seed pods with fertilizer and nutrients directly into the ground.

DroneSeed and BioCarbon are two such companies, both of which are developing drones that can carry a module that fires tree seeds into the ground at optimal locations. While currently designed for reforestation projects, it's not hard to imagine that the modules could be reconfigured to suit various agricultural seeds. With IoT and software for autonomous operation, a fleet of drones could complete extremely precise planting into the ideal conditions for growth of each crop, increasing the changes for faster growth and a higher crop yield.

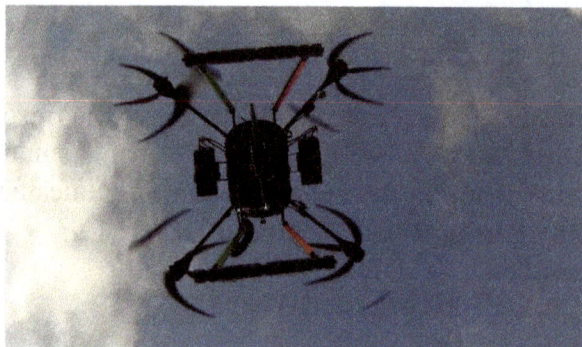

Example of a drone for tree planting.

Crop Spraying

There are also drones currently available and in development for crop spraying applications, offering the chance to automate yet another labor-intensive task. Using a combination of GPS, laser measurement and ultrasonic positioning, crop-spraying drones can adapt to altitude and location easily, adjusting for variables such as wind speed, topography and geography. This enables the drones to perform crop spraying tasks more efficiently, and with greater accuracy and less waste.

DJI Agras MG-1 crop spraying drone.

For example, DJI offers a drone called Agras MG-1 designed specifically for agricultural crop spraying, with a tank capacity of 2.6 gallons (10 liters) of liquid pesticide, herbicide or fertilizer, and a flight range of seven to ten acres per hour. Microwave radar enables this drone to maintain correct distance from the crops and ensure even coverage. According to DJI, it can operate automatically, semi-automatic or manual.

Working in conjunction with other agbots, crops identified as being in need of special attention could receive a personalized visit from the nearest drone at the first sign of trouble. Being able to provide individualized attention to any part of the field as soon as it's needed could help to stop many problems before they spread.

Agras MG-1 drone spraying a field.

Real-time Monitoring and Analysis

One of the most useful tasks drones can take on is remote monitoring and analysis of fields and crops. Imagine the benefits of using a small fleet of drones instead of a team of workers spending hours on their feet or in a vehicle travelling back and forth across the field to visually check crop conditions.

This is where the connected farm is essential, as all this data needs to be seen to be useful. Farmers can review the data, and only make personal trips out into the fields when there is a specific issue that needs their attention, rather than wasting time and effort by tending to healthy plants.

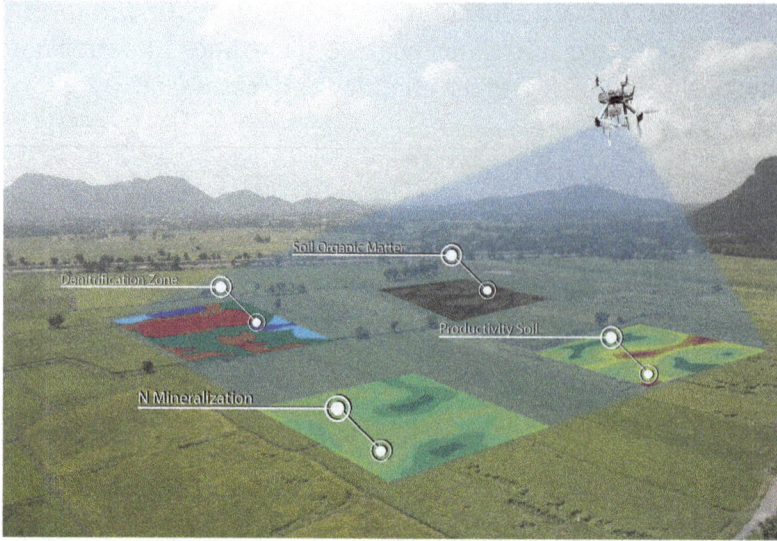

Given that drones for agricultural use are still early in their evolution, there are a few downsides. Ranges and flight times are not as robust as many farms would need—currently, even the longest running drones max out at around an hour of flight time before needing to return and recharge.

The capital expenses are also still quite high. Less expensive models exist, but they may not come with the necessary imaging or spraying equipment.

The Connected Farm: Sensors and the IoT

Innovative, autonomous agbots and drones are useful, but what will really make the future farm a "smart farm" will be what brings all this tech together: the Internet of Things.

The IoT has become a bit of a catch-all term for the idea of having computers, machines, equipment and devices of all types connected to each other, exchange data, and communicating in ways that enable them to operate as a so-called "smart" system. We're already seeing IoT technologies in use in many ways, such as smart home devices and digital assistants, smart factories and smart medical devices.

Smart farms will have sensors embedded throughout every stage of the farming process, and on every piece of equipment. Sensors set up across the fields will collect data on light levels, soil conditions, irrigation, air quality and weather. That data will go back to the farmer, or directly to AgBots in the field. Teams of robots will traverse the fields and work autonomously to respond to the needs of crops, and perform weeding, watering, pruning and harvesting functions guided by their own collection of sensors, navigation and crop data. Drones will tour the sky, getting the bird's eye view of plant health and soil conditions, or generating maps that will guide the robots, and help the human farmers to plan for the farm's next steps. All of this will help create higher crop production, and an increased availability and quality of food.

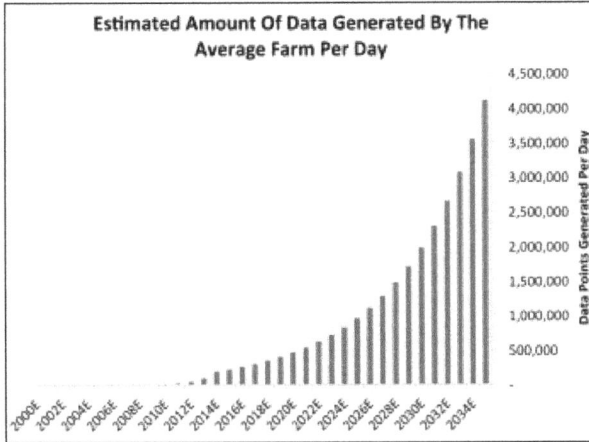

Estimated Amount Of Data Generated By The Average Farm Per Day

BI Intelligence shared their predictions that IoT devices installed in agriculture will increase from 30 million in 2015 up to 75 million by 2020. Under this trend, connected farms are expected to generate as many as 4.1 million data points each day in 2050—up from a mere 190,000 in 2014.

This mountain of data and other information generated by farming technology, and the connectivity enabling it to be shared, will be the backbone of the future smart farm. Farmers will be able to "see" all aspects of their operation—which plants are healthy or need attention, where a field needs water, what the harvesters are doing—and make informed decisions.

Automation is Changing Modern Farming

Agriculture in the modern age is changing rapidly. Rising global population and shifting trade policies affect the pricing, supply chain, and delivery of food products. Meanwhile, consumer preferences, especially in western countries, are shifting toward organic and sustainably-produced products and produce that require more attention, data, and labour.

These growing and changing demands need to be met by an agriculture industry that is facing labour shortages and rising costs for farm work. Many farmers are facing a dilemma between wanting to produce more, higher-quality crops and finding the workers to plant, maintain, and harvest those crops.

This tension is not new in agriculture. For all of human history, increasing agricultural production has been a function of either adding more labourers or finding more efficient tools to do the job. Modern agriculture is no different. In the face of labour shortages, farmers are turning to technology to make farms more efficient and automate the crop production cycle.

The growing interest in technology and automation is apparent in venture capital investments for ag tech startups.

These startups are addressing every aspect of the agriculture value chain. Some place remote sensors in the field to collect hyper-local data about growing conditions. Others create software to manage seed, soil, fertilizer, and irrigation, and make predictions about timing and yield. Some startups use drones to monitor conditions remotely and even apply fertilizers, pesticides and other treatments from above. A growing cohort of companies are working on agricultural robotics to build autonomous tractors, combines, and even fruit and vegetable picking robots.

Before you can automate farm operations, however, you need accurate data about the state of the farm. You also need a way for autonomous devices to connect with one another. This is the realm of the Internet of Things (IoT). IoT devices are the sensors, gauges and machines that are connected across a farm using Bluetooth, a cellular network, or some other type of connection. More IoT devices allow growers to collect more data about the state of their farms, and IoT is showing great promise for optimizing resource delivery and driving precision agriculture to achieve maximum efficiency.

For example, CropX is a company that installs soil sensors throughout the field to alert growers when soil conditions are outside the ordinary. A farmer might receive a notification that there is lower moisture levels in a certain part of the field. Given that information in real time, the farmer has an opportunity to correct the issue (in irrigation systems), to produce higher-quality and larger yields. Similar IoT sensor technology has applications in storage safety as well. OPI Systems' sensors for silos and elevators, for instance, track conditions and send alerts when heat or moisture might damage grain, or when a fire is possible.

It is not hard to see the value of small, connected devices throughout the farm for increasing efficiency and safety. It is no wonder growers have taken to IoT so strongly. According to Business Insider, IoT device installations in agricultural settings are expected to increase at a 20% annual compound rate over the next few years. With more accurate and timely data, growers can spend less time out in the field assessing and diagnosing conditions, and more time working on solutions.

The data from these IoT sensors, combined with information about seed, fertilizer, pesticide, rainfall and other factors, presents a compelling use case for software to optimize and predict growing conditions. Running a farm is a complex operation, with dozens of factors affecting every decision. As data collected from farms grows each year, software algorithms can become more complex in their predictive analytics, helping farmers decide when, where and how much irrigation, fertilizer, pesticide or other applications their farm needs.

Software can also automate the frustrating task of resource tracking and management. For instance, some software and apps enable growers to record all field applications and then track these resources from the field to multiple storage locations and then to the elevator. This resource management data can be incredibly powerful for optimizing agronomics on the farm, and it creates marketing value for food and beverage companies to differentiate themselves.

IoT and farm management software are currently available and growing in popularity and sophistication. The next big frontier in agricultural automation is machines and robots that can do the work for humans. While most of these applications are not fully commercially available or are too expensive currently for widespread use, many companies are racing to develop autonomous machines that are cost effective and reliably complete farm tasks throughout the growing cycle.

Perhaps the most iconic image of farm automation is the self-driving tractor. Since 1991, John Deere has been working on autonomy for its tractors. While many of its newer models have autonomous features, like line keeping and depth adjustments, all of Deere's models still require that a human sit inside. However, all major tractor manufacturers have plans and concepts in the works, for future full autonomy. Such a tractor that could be controlled remotely or even pre-programmed would provide significant savings in labour and input costs.

The DOT Power Platform.

There have been significant advancements in automating tasks on the farm by smaller innovative companies. DOT Technology Corp's autonomous, diesel-powered seeder follows predefined paths to sow an entire field without any human intervention. Rowbot is another company that has developed autonomous vehicles that can apply nitrogen between rows of corn and even sow cover crops under mature corn late in the season, before harvest.

In the UK, under human supervision, fully autonomous machinery planted, maintained and harvested a hectare of barley as part of the Hands Free Hectare project without a single human stepping in the field. While there were challenges with working in wet conditions and sacrificing some yield, the proof of concept project shows that it's theoretically possible to grow and harvest many commodities without human intervention.

Automation does not have to stop with tractors, and it is not limited to grains. Drones are gaining ground as autonomous vehicles that can provide information about the health of crops from above. They can quickly and cost effectively identify problem zones via imagery and infrared analysis and help farmers diagnose issues early on. Drones can also do some of the work typically left to airplanes or helicopters, dusting fields or even blowing water off ripe cherry trees after a heavy rainfall when the fruits are likely to burst.

Yamaha Precision Agriculture, for example, is testing a small, remote-piloted helicopter in Napa Valley that can spray fungicide. Vineyards in Napa often have tight row spacing and sometimes grow on steep hillsides. Until now, the only way to apply fungicide was for a farm worker to carry a backpack sprayer down the rows. Yamaha's helicopter can apply the fungicide from above, simplifying, improving, and speeding the application process.

Harvesting fruits and vegetables also proves to be a difficult problem to automate, but several companies are up to the challenge. Harvesting robots must be gentle with the produce to avoid bruising and with the plant to avoid damage. Abundant Robotics has developed an early version apple picking robot that is gentle and precise. Another company, Energid, has a similar solution for picking oranges. Neither of these solutions is ready for primetime and the complexity of a real orchard. Time will tell if they turn into a viable, cost effective solution to a difficult problem.

Every year, automation technology gets more sophisticated. What was cutting edge just a few years ago (guidance, drive-by-wire, continuously variable transmissions, remote sensors) has now become relatively commonplace and cost effective. Technological development toward full automation will only continue to accelerate as sensing, weather performance, terrain responsiveness and proactive autonomous decision making become more sophisticated.

It's not outside the realm of possibility that the coming decades could see farms managed from afar, with less need for human manual labour and more emphasis on human intuition, management and decision making. However, it is still clear that the human element of managing a farm is critical for the foreseeable future. Automation will enable farmers to scale their operations and be more efficient, but with all the complexities of weather and growing, it still takes human instincts and decision-making to run a modern farm.

Smart Irrigation Systems

The major problem in Agriculture with respect to irrigation is Rainfall and Water availability which has huge imbalance. Demand for water is increasing rapidly because of which there is need to save water and use the water efficiently to get productive yield, Smart Irrigation Systems comes into picture. The Water Level Controllers, Cyclic Timers, Auto Start Units for Submersible Pumps which control and monitor the activity of Irrigation plays a major role in saving water effectively.

Smart Sensors and Weather Tracking

Rapid growth in electronics and computer science has paved a major growth in Agricultural Automation. The Smart Sensors helps in recording the moisture and nitrogen levels which helps the farmers to choose perfect crop based on the soil fertility statistics and irrigate the land at the right time which avoids wrong investment on crops. Weather Tracking applications are booming in the Agricultural Market which helps in predicting the Rain, Wind Speed and Direction which helps in pre-planning and avoid loss of crop.

Farming and Robotics

Much like using robots and artificial intelligence in other industries, robotics within agriculture would improve productivity and would result in higher and faster yields. Such robots like the spraying and weeding robots recently acquired by John Deere can reduce agrochemical use by an incredible 90%. Other startup robotics companies are experimenting with laser and camera guidance for identifying and removing weeds without human intervention. These robots can use the guidance to navigate between rows of crops on its own, reducing the manpower behind it. Other companies are creating plant-transplanting robots that add a new level of efficiency to traditional methods and finally, automation is being tested for fruit-picking and nut harvesting, something that has always seemed to be too delicate for robotics in the past.

RFID Sensors and Tracking

After crops are harvested, RFID sensors can be used to track food from the field to the store. The end user, or the consumer, will be able to follow a detailed trail about the food they consume from the farm it came to the location where it was purchased. This technology could increase trustworthiness for manufacturers and their responsibility to provide fresh produce and goods.

Machine Learning and Analytics

Perhaps one of the most innovative pieces of the digital transformation is the ability to use machine learning and advanced analytics to mine data for trends. This can start way before the planting of the seed, with plant breeders. Machine learning can predict which traits and genes will be best for crop production, giving farmers all over the world the best breed for their location and climate.

Benefits of Automation in Agriculture

Site-specific agriculture requires the application of machinery equipped with high precision devices. Unfortunately, most performance specifications that must be met by machines or machine parts for use in precision agriculture can no longer be met through a traditional sequential design of the mechanism, the controllers, and the information systems. Increasingly, the improvement or adaptation of agricultural machines requires the application of a mechatronic design methodology to meet the stringent performance requirements that are essential for site-specific field operations. In a mechatronic design process, performance of the mechanism can be improved considerably or even optimized through the concurrent and integrated development of precision mechanisms, modern controllers, and advanced information systems.

Mass Flow Sensor for Combines

Sensor Requirements

During the past 15 years, research on yield sensors has focused mainly on the development of reliable grain flow sensors on combine harvesters for measuring the grain yield during harvesting. Although many sensors have been proposed, only a few proved to be suitable for commercial application due to the severe performance criteria imposed on the sensors, the most important of which are:

- The sensor should be able to measure the grain flow with sufficient accuracy such that measurement errors are less than 5%.

- Machine motion and vibration should not disturb the accuracy of the sensor.

- Analysis of the measurement signal before it becomes suitable for deriving yield maps should be simple and straightforward.

- The accuracy of the sensor must remain independent of variations in bulk properties.

- Requirements for recalibration and maintenance of the sensor should be limited.

- The sensor should have an appropriate design for easy integration in combines.

A yield sensor has been developed that amply meets the above-mentioned performance requirements.

Grain Yield Sensor

The proposed grain flow sensor is mounted at the outlet of the grain elevator, as shown in figure. The sensor consists of a 90-deg curved plate or chute, supported at the elevator housing by two pendulum rods that can rotate around a pivot point. A beam spring keeps the sensor in its initial position when the machine is at rest. A counterweight is fixed to the opposite tips of both rods such that the pivot point coincides with the center of gravity of the whole assembly so as to render the sensor insensitive to translational vibrations of the combine. In addition, this suspension drastically reduces the influence of driving uphill or downhill on the zero reading of the sensor. Normally, the threshed grain kernels are thrown by the pin parcels into the storage tank. To lead the grain flow smoothly into the sensor, a deflection plate and a rotor are installed at the head of the elevator.

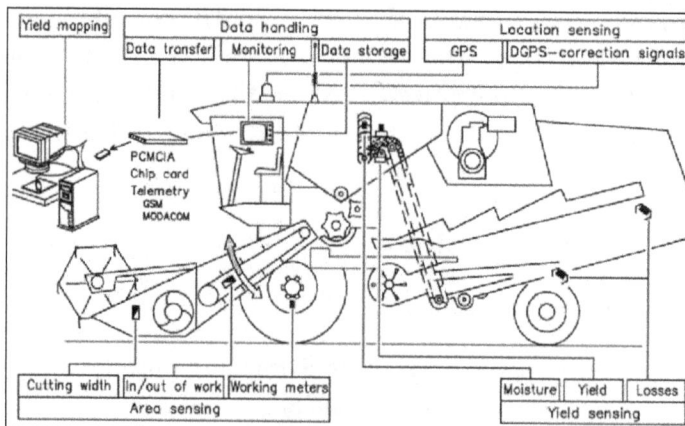

Grain elevator.

The grain mass flow entering the sensor exerts a force on the curved plate, causing the assembly to start rotating around its pivot point against the spring force. This final force is the result of the gravity force F_g, the centrifugal force F_{cp}, and the friction force F_f between the grain mass and the curved plate body and thus is a function of the total grain mass m on the plate. Consequently, the registered instantaneous deflection of the beam spring by a linear inductive distance sensor is a measure of the mass flow variations in the curved plate.

Unfortunately, the friction coefficient in the friction force is function of kernel characteristics such as crop type, and moisture content. As a consequence, the sensor must be re-calibrated for different crops and varying harvesting conditions, a very time- consuming and delicate operation. However by properly selecting the projection angle or the direction in which the force on the chute is measured, the effect of the friction coefficient m can be minimized such that the grain yield sensor is almost independent of the friction parameters of the kernels. After optimization, the influence of friction is less than 0.5 % per 0.1 change of the friction coefficient. For normal grain, the friction coefficient m varies between 0.1 and 0.7.

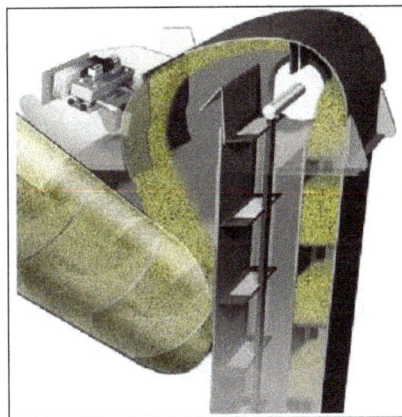

Grain flow sensor.

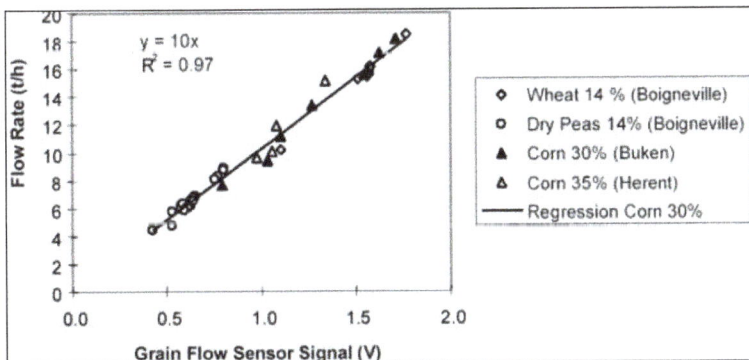

Sensor calibration (the sensor characteristics are independent of crop type and crop condition).

The optimized grain flow sensor has been tested over several years under widely varying harvesting conditions ranging from winter barley with a moisture content of 12% to corn with a moisture content of 40%. The regression lines in figure show that the sensor is independent of crop type and crop condition. Only the harvesting season influences the slope of the regression line, indicating that the sensor should be calibrated once a year at the star of the harvesting season.

The accuracy of the grain yield sensor was evaluated on harvested areas of various sizes ranging from 120 to 2000 m². The registered yield error is due to inaccuracies in the measurement procedure and sensor inaccuracies. The error in percentage yield increases with decreasing harvested area. For a harvested area of 400 m², matching the grid size of 20 m x 20 m for soil sampling, the maximum error was 5%. For an area of 2000 m², the maximum error decreased to 3%. The error in estimating the yield of a 6-ha field in the Netherlands was less then 1.8%.

Grain Yield Maps

To transform the mass flow rate data from the yield sensor into a yield map, additional information is collected by the following sensors:

- A capacitive moisture sensor is mounted in the grain elevator to convert the mass flow rate measured at a certain moisture content into a mass flow rate with a standard moisture content (e.g., 14%).

- As a larger cutting width directly influences the mass flow rate in the curved plate, an ultrasonic distance sensor is installed on the header of the combine to measure the cutting width of the knife, which influences the mass.

- A precise Doppler radar sensor to measure the travel speed of the combine, in combination with the ultrasonic sensor outputs, is necessary to relate the actual harvested surface to the measured grain mass flow in the chute.

- To relate the grain yield to the correct location in the field, the absolute position of the combine is determined by a Differential Global Positioning System DGPS).

- The transportation time the grain kernels need to reach the yield sensor after the crop is cut by the cutter bar and the smearing effect of the return loop where unthreshed ears are brought back into the threshing process should be compensated in the yield measurements. To this end, an analytical model of the grain flow process has been developed in the New Holland TF78 combine. This model starts by representing the biomass flow above the cutter bar. Subsequently, it describes the transport time of the biomass through the feeding auger and the transport time of the unthreshed kernels in the threshing-sieving mechanism. Once the grain has fallen through the concaves of the threshing drums, the kernel distribution and transport time on the grain pan and the sieves is modeled. A similar model is provided for the return loop. In a final step, the residence time of the kernels in the grain elevator before reaching the yield sensor is modeled.

Site-specific Spraying

Chemical Crop Protection

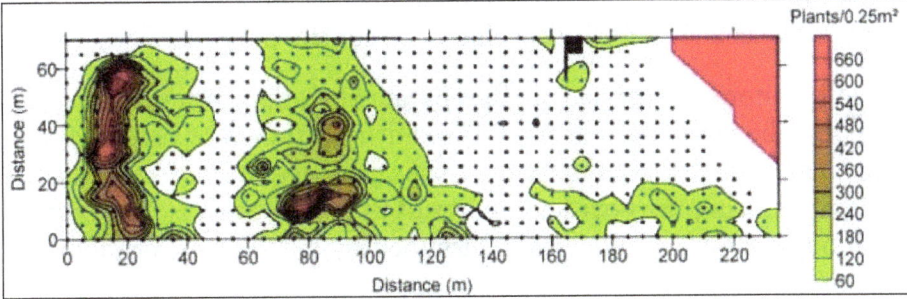

Typical pattern of development of weed populations in the early stage of infestation.

Agricultural production suffers from severe losses due to insects, plant diseases, and weeds. Owing to an exponentially growing world population, crop protection has become one of the most important field operations to increase productivity and crop yield. The most widely used practice in weed control is spraying herbicides uniformly over the agricultural fields at various times during the cultivation cycle of arable crops. To guarantee their effectiveness, over application of pesticides is commonly advised; however, excessive use of pesticides raises the danger of toxic residue levels on agricultural products. Because pesticides, and especially herbicides, are a major cost factor in the production of field crops and have been identified as a major contributor to ground water and surface water contamination, their use must be reduced dramatically.

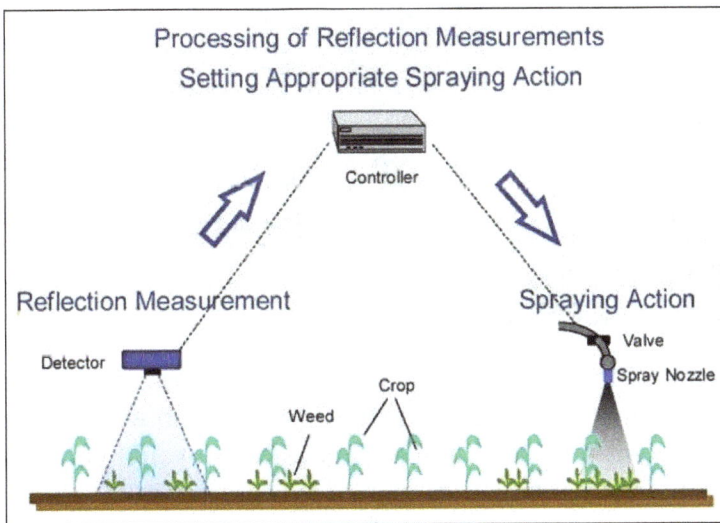

Weed-activated spraying.

Fortunately, most weed populations develop in patches in the field, with large areas of the field remaining free of weeds or having a very low weed density in the early stage of infestation. As a consequence, herbicides would be used more efficiently if they were applied in the appropriate dose, where they are needed, and not to areas with insignificant weed densities. Thus, weeds have been suggested as the primary target for spatially selective pest control.

To set up a local weed treatment, the weed populations must be evaluated in the field. In this respect, two concepts of site-specific weed control have been suggested:

- Weed monitoring is carried out in separate operations prior to the spraying operation ("the mapping concept"). Weed distribution is represented in digitized weed maps, that are later used during spraying operations to activate the spraying system using the board computer of the field sprayers. The instantaneous position of the field vehicle is determined by a GPS receiver mounted on the machine.

- Weed monitoring and spraying are carried out sequentially in the same operation ("the real-time concept"). A real-time weed detection system mounted on the field-spraying machine detects "individual" weeds and transmits that information to a control system that controls the spraying equipment of the vehicle. This is called weed-activated spraying.

Artificial Intelligence in Agriculture

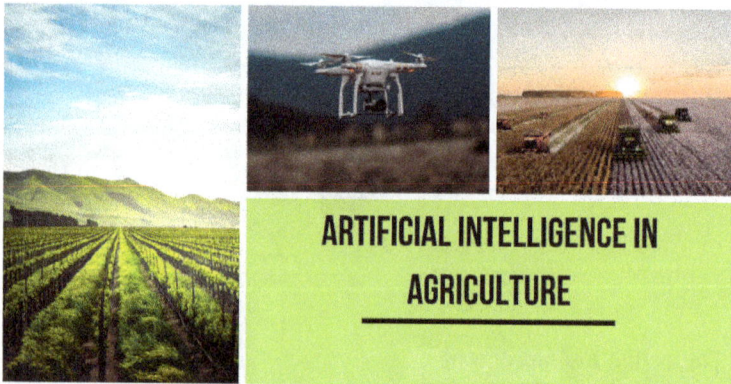

ARTIFICIAL INTELLIGENCE IN AGRICULTURE

The world of Artificial Intelligence (AI) is quickly on the rise as it makes its way into many different industries. From manufacturing to automotive, you can likely see AI used for many different purposes; and as time goes on, you'll only see it even more. One of the most interesting industries that AI is breaking into is agriculture. Agriculture is a major industry and a huge part of the foundation of our economy. According to the Environmental Protection Agency (EPA), the agricultural industry contributes nearly $330 billion in annual revenue to our economy. As climates are changing and populations are increasing, AI is becoming a technological innovation that is improving and protecting crop yield in the U.S. Here are some major ways that AI is contributing to the agricultural industry:

Robotics

AI companies are focusing much of their efforts on developing autonomous robots that can easily handle multiple agricultural tasks. These robots are capable of harvesting crops at a much faster pace and higher volume than human workers. The robots are designed to assist in picking and packing crops while also combating other challenges within the agricultural labor force. Additionally, agricultural robots have the ability to protect crops from harmful weeds that may be resistant to herbicide chemicals that are meant to eliminate them.

Image-based Insights for Crop/Soil Health

Many farms are taking advantage of drone technology to provide high-quality imaging that can help monitor crops while scanning and analyzing fields to collect necessary agricultural data. This imaging technology can also assist in the identification of crops and their progress, including their health, and the determination of their readiness. For example, these images can provide farmers with the ability to determine how ripe their crops are, and if and when they will be ready for harvest. Additionally, imaging technology can assist with overall field management, providing estimates in real-time identifying where specific crops may require more water, fertilizer, soil or pesticides. Machine learning is also used to provide an analysis of crop or soil health. Innovative AI and machine learning companies have developed technologies that use machine learning to provide farmers and laborers with insights about the strengths and weaknesses of their soil. This is done with the intention of preventing and eliminating bad crops and increasing the potential for healthy crops to grow.

Precise Farming

Precision farming uses AI to generate accurate and controlled techniques that help provide guidance and understanding about water and nutrient management, optimal harvesting and planting times as well as when the right times for crop rotation would be. These processes make farming more efficient, and can even help predict ROI on specific crops based on their costs and margin within the market.

While AI won't eliminate the jobs of human farmers, it will surely improve their processes and provide them with more efficient ways to produce, harvest and sell essential crops.

Sowing the Seeds

High-tech agriculture starts at the very second that the seed is first placed in the ground.

Experts in the field are familiar with "variable rate planting equipment" that does more than just planting a seed down into the dirt somewhere.

All sorts of artificial intelligence work is being done behind the scenes on predictions — where a seed will grow best, what soil conditions are likely to be, etc. The power of artificial intelligence is being applied to agricultural big data in order to make farming much more efficient — and that's only the beginning.

Perhaps a better question would be "What's picking your food?" That's right — companies are already producing robotic harvesting equipment, partially in response to labor gaps that have left farmers scrambling to harvest crops like fruits and berries.

You can see a lot of this at work in documentation from companies like Harvest Croo, which has produced an autonomous strawberry picking machine, and Abundant Robotics, where a vacuum apparatus harvests mature apples from trees.

While manufacturing robots have been around for a while, these harvesting technologies are really something new in many ways. Harvest technologies like the Harvest Croo berry picker operate on the basis of machine vision and sensor fusion to "see" where harvest fruits and berries are. They use sophisticated directed movements to pick precisely. This is the kind of functionality that is very much in the "artificial intelligence" field and mimics human cognition and directed action.

Agricultural robotics is filling a need as labor pools decrease. But it is also saving humans from one of the most repetitive and difficult jobs in our economy.

People simply do not want to have to harvest enormous amounts of produce all day, every day. In that sense, harvest robotics is also making our world a little more enjoyable. Now, if people do want to pick by hand, boutique orchards and gardens offer that option.

Eye in the Sky

How are farms using artificial intelligence to direct crop planting, harvesting and more, and how are they getting that data in the first place?

In some trade journals, you can see unmanned aerial vehicles or drones being outfitted with precision sensors, in order to run the fields and get the data that's needed. These airborne surveillance engines can look for stunted crops, signs of pest or weed damage, dryness and many other variables that are part of the difficulty of farming in general. With all of this data in hand, farmers can enhance their production models and their strategies across the lay of the land to decrease risk, waste and liability.

Pest and Weed Control

Yesterday's farmers were living in fear of the windstorm and the grasshopper — not anymore. Farmers are quickly adopting new high-tech ways of protecting plants against weeds and various kinds of pests outdoors. Another alternative is to grow in greenhouses, which is being done as well, but some of the most amazing farming technology is being deployed outside.

The "See and Spray" model acquired by John Deere recently is an excellent example of harnessing the power of artificial intelligence and computer vision.

"We welcome the opportunity to work with a Blue River Technology team that is highly skilled and intensely dedicated to rapidly advancing the implementation of machine learning in agriculture," John May, president, agricultural solutions, and chief information officer at Deere, said in a press statement, weighing in on the potential of new technologies in farming. "As a leader in precision agriculture, John Deere recognizes the importance of technology to our customers. Machine learning is an important capability for Deere's future.

We know that artificial intelligence excels at image processing — computers can now "see" almost as well as we can. So by deploying mobile technologies with AI and computer vision built in, farmers can find weeds and eradicate them, instead of blanket spraying an entire crop. That makes the food cleaner, and it saves enormous amounts of money. It is just another example of real new technologies that are having a dramatic impact on yields and everything else.

Yield Boosting Algorithms

When we talk about machine learning and artificial intelligence, we often talk about algorithms. The mathematical models behind computer science are the fundamental basis for how we deal with big data to make decisions.

Companies are now quickly developing agricultural yield boosting algorithms that can show farmers what's going to be best for a crop. Despite some concerns about the difficulty of doing this type of analysis in nature, farmers and others have been able to make quite a lot of headway in maximizing crop yield, simply by applying the algorithms and intelligence generators that we've built to help computers imitate our own cognitive abilities.

The Farmer's Alexa

There is one more very interesting groundbreaking technology that might also be one of the highlights in the modern farmer's tech toolkit.

Imagine a tired farmer sitting down to dinner at the end of a work day, puzzling over some conundrum — how to keep the crows out of the corn, or whether to seed a rocky patch of earth.

Propping his head on his arms, he directs his question to the next room:

"Alexa?"

Yes, companies are talking about creating chatbots for farmers, artificial intelligence personalities like the smart home helper "Alexa" that are able to converse with farmers to help them figure out tough problems. We're hoping that these specialized farming chatbots are a little more capable than Alexa, since the current consumer technology basically provides encyclopedic facts and figures and not much else. However, if they're packed with the right answers and analytics information, the farmer's chatbot could be a real boon to busy farm managers who are doing all they can to expand and grow their businesses.

These are some of the best new technologies coming out to help farmers produce all the food that we need in a rapidly changing world. Population growth and climate change will be massive challenges, but artificial intelligence deployment can help blunt the impact of these and other challenges, and make smart farming much more resistant to the problems that farmers face.

References

- Automation-new-horizon-in-agricultural-machinery, open-access: longdom.org, Retrieved 02 January, 2019

- Smart-FarmingAutomated-and-Connected-Agriculture- 16653: engineering.com, Retrieved 23 March, 2019

- Automation-is-changing-modern-farming: realagriculture.com, Retrieved 19 May, 2019

- Latest-agriculture-automation-trends-in-2019: narenelectrix.in, Retrieved 03 July, 2019

- Artificial-intelligence-in-agriculture, datadriveninvestor: medium.com, Retrieved 14 August, 2019

- The-6-most-amazing-ai-advances-in-agriculture- 33177: techopedia.com, Retrieved 25 January, 2019

Precision Agriculture

Precision Agriculture is a type of farming management which is based upon observation and analysis of inter and intra-field variability in crops. It uses different tools and technologies such as GPS, sensors, variable rate seeding, weather modeling, etc. The topics elaborated in this chapter will help in gaining a better perspective about these tools and related aspects of precision agriculture.

Precision agriculture (PA) is an approach to farm management that uses information technology (IT) to ensure that the crops and soil receive exactly what they need for optimum health and productivity. The goal of PA is to ensure profitability, sustainability and protection of the environment. PA is also known as satellite agriculture, as-needed farming and site-specific crop management (SSCM).

Precision agriculture relies upon specialized equipment, software and IT services. The approach includes accessing real-time data about the conditions of the crops, soil and ambient air, along with other relevant information such as hyper-local weather predictions, labor costs and equipment availability. Predictive analytics software uses the data to provide farmers with guidance about crop rotation, optimal planting times, harvesting times and soil management.

Sensors in fields measure the moisture content and temperature of the soil and surrounding air. Satellites and robotic drones provide farmers with real-time images of individual plants. Information from those images can be processed and integrated with sensor and other data to yield guidance for immediate and future decisions, such as precisely what fields to water and when or where to plant a particular crop.

Agricultural control centers integrate sensor data and imaging input with other data, providing farmers with the ability to identify fields that require treatment and determine the optimum amount of water, fertilizers and pesticides to apply. This helps the farmer avoid wasting resources and prevent run-off, ensuring that the soil has just the right amount of additives for optimum health, while also reducing costs and controlling the farm's environmental impact.

In the past, precision agriculture was limited to larger operations which could support the IT infrastructure and other technology resources required to fully implement and benefit from the benefits of precision agriculture. Today, however, mobile apps, smart sensors, drones and cloud computing makes precision agriculture possible for farming.

Precision agriculture merges the new technologies borne of the information age with a mature agricultural industry. It is an integrated crop management system that attempts to match the kind and amount of inputs with the actual crop needs for small areas within a farm field. This goal is not new, but new technologies now available allow the concept of precision agriculture to be realized in a practical production setting.

Precision agriculture often has been defined by the technologies that enable it and is often referred to as GPS (Global Positioning System) agriculture or variable-rate farming. As important as the devices are, it only takes a little reflection to realize that information is the key ingredient for precise farming. Managers who effectively use information earn higher returns than those who do not.

Precision farming distinguishes itself from traditional agriculture by its level of management. Instead of managing whole fields as a single unit, management is customized for small areas within fields. This increased level of management emphasizes the need for sound agronomic practices. Before considering the jump to precision agriculture management, a good farm management system must already be in place.

Precision agriculture is a systems approach to farming. To be viable, both economic and environmental benefits must be considered, as well as the practical questions of field-level management and the needed alliances to provide the infrastructure for technologies. The issues surrounding precision agriculture include perceived benefits and also barriers to widespread adoption of precision agriculture management.

Management

- Data acquistion and analysis.
- Decision support system.
- Increased attention to management.
- Learning curve.

Economics

- Changes in costs.
- Changes in revenues.
- Cash flow.
- Risk.

Alliances

- Accurate GPS availability.

- Variable rate technology availability.

- Site-specific managment services availability.

- Financing.

Environmental

- Decrease input losses.

- Target nutrients to increase uptake efficiency.

The Need for Precision Agriculture

Farmers usually are aware that their fields have variable yields across the landscape. These variations can be traced to management practices, soil properties and/or environmental characteristics. Soil characteristics that affect yields include texture, structure, moisture, organic matter, nutrient status and landscape position. Environmental characteristics include weather, weeds, insects and disease.

The aerial photo in figure illustrates that in some fields, within-field variability can be substantial. In this field, the best crop growth was near waterways and level areas of the field. Sideslopes where erosion depleted topsoil showed moisture stress and reduced plant stands. the variation in yield levels for corn and soybean is typically 2 to 1.

Aerial view of a central Missouri field.

Seeing this magnitude of variation prompts most farmers to ask how the problem that is causing the low yields can be fixed. There is no economically feasible method of "fixing" the depleted topsoil areas in this field, so the management challenge is to optimally manage the areas within the field that have different production capacities. This does not necessarily mean having the same yield level in all areas of the field.

A farmer's mental information database about how to treat different areas in a field required years of observation and implementation through trial-and-error. Today, that level of knowledge of field conditions is difficult to maintain because of the larger farm sizes and changes in areas farmed due to annual shifts in leasing arrangements. Precision agriculture offers the potential to automate and simplify the collection and analysis of information. It allows management decisions to be made and quickly implemented on small areas within larger fields.

Tools of Precision Agriculture

In order to gather and use information effectively, it is important for anyone considering precision farming to be familiar with the technological tools available. These tools include hardware, software and recommended practices.

Global Positioning System (GPS) Receivers

Global Positioning System satellites broadcast signals that allow GPS receivers to calculate their position. This information is provided in real time, meaning that continuous position information is provided while in motion. Having precise location information at any time allows soil and crop measurements to be mapped. GPS receivers, either carried to the field or mounted on implements allow users to return to specific locations to sample or treat those areas.

Uncorrected GPS signals have an accuracy of about 300 feet. To be useful in agriculture, the uncorrected GPS signals must be compared to a land-based or satellite-based signal that provides a position correction called a *differential* correction. The corrected position accuracy is typically 63-10 feet. In Missouri, the Coast Guard provides differential correction beacons that are available to most areas free of charge. When purchasing a GPS receiver, the type of differential correction and its coverage relative to use area should be considered.

Yield Monitoring and Mapping

Grain yield monitors continuously measure and record the flow of grain in the clean-grain elevator of a combine. When linked with a GPS receiver, yield monitors can provide data necessary for yield maps. Yield measurements are essential for making sound management decisions. However, soil, landscape and other environmental factors should also be weighed when interpreting a yield map. Used properly, yield information provides important feedback in determining the effects of managed inputs such as fertilizer, lime, seed, pesticides and cultural practices including tillage and irrigation.

Yield measurements from a single year may be heavily influenced by weather. Examining yield information records from several years and including data from extreme weather years helps in determining if the observed yield level is due to management or is climate-induced.

Grid Soil Sampling and Variable-rate Fertilizer (VRT) Application

The recommended soil sampling procedure is to take samples from portions of fields that are no more than 20 acres in area. Soil cores taken from random locations in the sampling area are combined and sent to a laboratory to be tested. Crop advisors make fertilizer application recommendations from the soil test information for the 20-acre area. Grid soil sampling uses the same principles of soil sampling but increases the intensity of sampling. For example, a 20-acre sampling area would have 10 samples using a 2-acre grid sampling system (samples are spaced 300 feet from each other) compared to one sample in the traditional recommendations. Soil samples collected in a systematic grid also have location information that allows the data to be mapped. The goal of grid soil sampling is a map of nutrient needs, called an application map. Grid soil samples are analyzed in the laboratory, and an interpretation of crop nutrient needs is made for each soil sample. Then the fertilizer application map is plotted using the entire set of soil samples. The application map is loaded into a computer mounted on a variable-rate fertilizer spreader. The computer uses the application map and a GPS receiver to direct a product-delivery controller that changes the amount and/or kind of fertilizer product, according to the application map.

Remote Sensing

Remote sensing is collection of data from a distance. Data sensors can simply be hand-held devices, mounted on aircraft or satellite-based. Remotely-sensed data provide a tool for evaluating crop health. Plant stress related to moisture, nutrients, compaction, crop diseases and other plant health concerns are often easily detected in overhead images. Electronic cameras can also record near-infrared images that are highly correlated with healthy plant tissue. New image sensors with high spectral resolution are increasing the information collected from satellites.

Remote sensing can reveal in-season variability that affects crop yield, and can be timely enough to make management decisions that improve profitability for the current crop. Remotely-sensed images can help determine the location and extent of crop stress. Analysis of such images used in tandem with scouting can help determine the cause of certain components of crop stress. The images can then be used to develop and implement a spot treatment plan that optimizes the use of agricultural chemicals.

Crop Scouting

In-season observations of crop conditions may include:

- Weed patches (weed type and intensity).

- Insect or fungal infestation (species and intensity).

- Crop tissue nutrient status.

- Flooded and eroded areas.

Using a GPS receiver on an all-terrain vehicle or in a backpack, a location can be associated with observations, making it easier to return to the same location for treatment. These observations also can be helpful later when explaining variations in yield maps.

Geographic Information Systems (GIS)

Geographic information systems (GIS) are computer hardware and software that use feature attributes and location data to produce maps. An important function of an agricultural GIS is to store layers of information, such as yields, soil survey maps, remotely sensed data, crop scouting reports and soil nutrient levels. Geographically referenced data can be displayed in the GIS, adding a visual perspective for interpretation.

In addition to data storage and display, the GIS can be used to evaluate present and alternative management by combining and manipulating data layers to produce an analysis of management scenarios.

Information Management

The adoption of precision agriculture requires the joint development of management skills and pertinent information databases. Effectively using information requires a farmer to have a clear idea of the business' objectives and of the crucial information necessary to make decisions. Effective information management requires more than record-keeping analysis tools or a GIS. It requires an entrepreneurial attitude toward education and experimentation.

Identifying a Precision Agriculture Service Provider

Farmers should consider the availability of custom services when making decisions about adopting site-specific crop management. Agricultural service providers may offer a variety of precision agriculture services to farmers. By distributing capital costs for specialized equipment over more land and by using the skills of precision agriculture specialists, custom services can decrease the cost and increase the efficiency of precision agriculture activities.

The most common custom services that precision agriculture service providers offer are intensive soil sampling, mapping and variable rate applications of fertilizer and lime. Equipment required for these operations include a vehicle equipped with a GPS receiver and a field computer for soil sampling, a computer with mapping software and a variable-rate applicator for fertilizers and lime. Purchasing this equipment and learning the necessary skills is a significant up-front cost that can be prohibitive for many farmers.

Agricultural service providers must identify a group of committed customers to justify purchasing the equipment and allocating human resources to offer these services. Once a service provider is established, precision agriculture activities in that region tend to center around the service providers. For this reason, adopters of precision farming practices often are found in clusters surrounding the service provider.

Main Characteristics of Precision Farming

Every agricultural intervention inevitably presumes an accurate knowledge of the arable sites. In the 1980s the so called industrialized agricultural practice organised the farms into producing blocks, taking into account the heterogeneity of arable sites only partly despite of the existing technical facilities. Although the yields were very high, they were achieved by high energy inputs (fuel, fertilizers, plant protection agents etc.) and by low efficiency. Materials that were not used up by the agro-ecosystem endangered the environment potentially. Energy and environmental crisis, reducing agricultural efficiency, reducing supports, as well as the increasing Earth's population and the rapid growth of starvation revealed that agriculture is in a global crisis.

Information-technology (IT) and its spread was a real breakthrough in handling the crisis. The utilization of the information technology (IT) in farming happened in the form of precision agriculture. The principle of precision farming is to connect geo-positioning systems with traditional farming practice.

In plant growing precision farming includes:

- Remote sensing,
- Utilization of data gained by remote sensing with the help of geo-positioning systems,
- Species and variety-specific sowing,
  - Plant number (plant width, row-width),
  - Adjustment of the sowing depth.
- Plant-tending based on survey,
- Nutrient replacement determined by the nutrient reserves and the actual state of plant development,
- Integrated plant protection,
- Yield-modelling,
- Preparation of statistical analysis.

The most important differences in traditional and precision farming are:

Traditional farming	Precision farming
Unit of treatment and organisation: the Meld that is regarded as a homogenous arable site.	Unit of treatment and organisation: arable site that is regarded as different from one point to the other and at "field level as heterogeneous.
Nutrient management based on average sample taking.	Nutrient management based on GPS and point-like sample taking.
Average survey. on plant deceases and damage and intervention if necessary.	Plant protection treatments based on GPS and point-like plant survey.
Sowing with same plant number and variety.	Plant species and plant variety-specific sowing.
Same machine operation practice.	Machine-operation adjusted to the arable Site.
Unified plant stock in space and time.	Unified plant stock organised into homogeneous blocks at arable sites.
Few data influencing decision preparation.	A lot of data influencing decision preparation.

In traditional farming practice the field is the smallest unit. Through spreading satellite GPS-systems we are able to determine our position on a given site at any time and continuously. As a result different plant growing treatments can be carried out on sites smaller than a field-size. E.g., it is possible to consider the differing physical and chemical soil conditions on a given site and to trace weed infestation and the damages caused by pests and pathogens within the field.

The available computer systems and as a result the collected data base by geo-information systems (GIS: Geographic Information System) enable us to make an overall picture about our agricultural area and we can make well supported economic decisions based on the collected information. Decision-making helps us to realize agro-technical treatments adjusted to the differences of the given arable site.

Plant protection, producing quality products, protection of our arable sites and last but not at least lower costs and higher economic efficiency support the use of precision farming in future widely. With the use of precision farming we can increase the intensity of production and reduce environmental pressure and at the same time we can increase the quality at lower costs as well. Applying the system on larger areas – on several thousand hectares – is also justified, because it can reduce the specific costs. It is obvious, that precision farming is one of the basic tools of sustainable farming nowadays.

Remote sensing is the acquisition of information about an object in our environment (mostly about the surface of the Earth) without making any physical contact with the object or phenomenon. Further more we should mention the analysis of the collected data together with professional, technical background. Remote sensing spread by using areal sensor technologies on satellites. This is one of the most efficient method of data collection.

Remote Sensing Applications

Crop production forecasting: Remote sensing is used to forecast the expected crop production and yield over a given area and determine how much of the crop will be harvested under specific conditions. Researchers can be able to predict the quantity of crop that will be produced in a given farmland over a given period of time.

Assessment of crop damage and crop progress: In the event of crop damage or crop progress, remote sensing technology can be used to penetrate the farmland and determine exactly how much of a given crop has been damaged and the progress of the remaining crop in the farm.

Horticulture, cropping systems analysis: Remote sensing technology has also been instrumental in the analysis of various crop planting systems. This technology has mainly been in use in the horticulture industry where flower growth patterns can be analyzed and a prediction made out of the analysis.

Crop identification: Remote sensing has also played an important role in crop identification especially in cases where the crop under observation is mysterious or shows some mysterious characteristics. The data from the crop is collected and taken to the labs where various aspects of the crop including the crop culture are studied.

Crop acreage estimation: Remote sensing has also played a very important role in the estimation of the farmland on which a crop has been planted. This is usually a cumbersome procedure if it is carried out manually because of the vast sizes of the lands being estimated.

Crop condition assessment and stress detection: Remote sensing technology plays an important role in the assessment of the health condition of each crop and the extent to which the crop has withstood stress. This data is then used to determine the quality of the crop.

Identification of planting and harvesting dates: Because of the predictive nature of the remote sensing technology, farmers can now use remote sensing to observe a variety of factors including the weather patterns and the soil types to predict the planting and harvesting seasons of each crop.

Crop yield modelling and estimation: Remote sensing also allows farmers and experts to predict the expected crop yield from a given farmland by estimating the quality of the crop and the extent of the farmland. This is then used to determine the overall expected yield of the crop.

Identification of pests and disease infestation: Remote sensing technology also plays a significant role in the identification of pests in farmland and gives data on the right pests control mechanism to be used to get rid of the pests and diseases on the farm.

Soil moisture estimation: Soil moisture can be difficult to measure without the help of remote sensing technology. Remote sensing gives the soil moisture data and helps in determining the quantity of moisture in the soil and hence the type of crop that can be grown in the soil.

Irrigation monitoring and management: Remote sensing gives information on the moisture quantity of soils. This information is used to determine whether a particular soil is moisture deficient or not and helps in planning the irrigation needs of the soil.

Soil mapping: Soil mapping is one of the most common yet most important uses of remote sensing. Through soil mapping, farmers are able to tell what soils are ideal for which crops and what soil require irrigation and which ones do not. This information helps in precision agriculture.

Monitoring of droughts: Remote sensing technology is used to monitor the weather patterns including the drought patterns over a given area. The information can be used to predict the rainfall patterns of an area and also tell the time difference between the current rainfall and the next rainfall which helps to keep track of the drought.

Land cover and land degradation mapping: Remote sensing has been used by experts to map out the land cover of a given area. Experts can now tell what areas of the land have been degraded and which areas are still intact. This also helps them in implementing measures to curb land degradation.

Identification of problematic soils: Remote sensing has also played a very important role in the identification of problematic soils that have a problem in sustaining optimum crop yield throughout a planting season.

Crop nutrient deficiency detection: Remote sensing technology has also helped farmers and other agricultural experts to determine the extent of crop nutrients deficiency and come up with remedies that would increase the nutrients level in crops hence increasing the overall crop yield.

Reflectance modeling: Remote sensing technology is just about the only technology that can provide data on crop reflectance. Crop reflectance will depend on the amount of moisture in the soil and the nutrients in the crop which may also have a significant impact on the overall crop yield.

Determination of water content of field crops: Apart from determining the soil moisture content, remote sensing also plays an important role in the estimation of the water content in the field crops.

Crop yield forecasting: Remote sensing technology can give accurate estimates of the expected crop yield in a planting season using various crop information such as the crop quality, the moisture level in the soil and in the crop and the crop cover of the land. When all of this data is combined it gives almost accurate estimates of the crop yield.

Flood mapping and monitoring: Using remote sensing technology, farmers and agricultural experts can be able to map out the areas that are likely to be hit by floods and the areas that lack proper drainage. This data can then be used to avert any flood disaster in future.

Collection of past and current weather data: Remote sensing technology is ideal for collection and storing of past and current weather data which can be used for future decision making and prediction.

Crop intensification: Remote sensing can be used for crop intensification that includes collection of important crop data such as the cropping pattern, crop rotation needs and crop diversity over a given soil.

Water resources mapping: Remote sensing is instrumental in the mapping of water resources that can be used for agriculture over a given farmland. Through remote sensing, farmers can tell what water resources are available for use over a given land and whether the resources are adequate.

Precision farming: Remote sensing has played a very vital role in precision agriculture. Precision agriculture has resulted in the cultivation of healthy crops that guarantees farmers optimum harvests over a given period of time.

Climate change monitoring: Remote sensing technology is important in monitoring of climate change and keeping track of the climatic conditions which play an important role in the determination of what crops can be grown where.

Compliance monitoring: For the agricultural experts and other farmers, remote sensing is important in keeping track of the farming practices by all farmers and ensuring compliance by all farmers. This helps in ensuring that all farmers follow the correct procedures when planting and when harvesting crops.

Soil management practices: Remote sensing technology is important in the determination of soil management practices based on the data collected from the farms.

Air moisture estimation: Remote sensing technology is used in the estimation of air moisture which determines the humidity of the area. The level of humidity determines the type of crops to be grown within the area.

Crop health analysis: Remote sensing technology plays an important role in the analysis of crop health which determines the overall crop yield.

Land mapping: Remote sensing helps in mapping land for use for various purposes such as crop growing and landscaping. The mapping technology used helps in precision agriculture where specific land soils are used for specific purposes.

Sample-taking Strategies

The principle of site-specific sample taking includes marking the sites of sample-taking within the field of observation, where we do the measurements and the data of measurement (as reliable as possible) give us conclusions about the characteristics of the total area.

Sample-taking strategies can be as follows:

Traditional (randomly arranged) sample-taking: Traditional sample-taking means soil-sampling per 5 hectares, possibly not on one site, but samples were taken on site randomly arranged. The samples taken this way are mixed and the unified sample is analyzed, giving the base for the homogeneous treatment of the field. This type of sampling strategy does not enable us to consider the diversity of the field, there is a great chance to make a random failure and we cannot evaluate the changes between the times of soil-sampling therefore this method of sampling cannot be applied by precision plant growing.

Sample-taking alongside typical (management) zones: This type of sampling takes the soil samples from area units marked by former experiences. Within the marked zones soil samples can be taken randomly, because they do not influence the values of the aggregated samples with alternating data. Management zones cover the whole territory of the field and make a precision treatment possible.

Sample taking according to chosen typical areas: Based on former experience this type of methods marks individual areas on the field (e.g., based on soil maps or former yield-data), or direct sampling is also possible based on satellite images. It returns to these marked areas nearly every time when taking soil samples. This way we can trace the changes on the same part of the field. A disadvantage of the method is that it does not offer a complete image of the field and very often precision treatments cannot be carried out accurately. An advantage is that it is time and cost efficient.

Sample-taking alongside a grid: Grid-sampling divides the field into treatment units and presumes that parts have similar soil characteristics within the treatment unit. There is no doubt that this method is labour and cost demanding, but it is necessary to site-specific treatments. A great advantage is that data gained between the sample taking times can be compared and can be evaluated according to different aspects (state of nutrient supply, environment protection etc.). Grid-sampling covers the field totally and no uncertain parts remain during the treatments.

Some sub-types of grid-sampling:

- Within the grid randomly.

- Within the grid diagonally.

- In the centre of the area covered by the grid randomly.

- In systematic points.

Tools and Equipments

Precision farming is a combination of application of different technologies. All these combinations are mutually inter related and responsible for developments.

Global positioning system (GPS): It is a set of 24 satellites in the Earth orbit. It sends out radio signals that can be processed by a ground receiver to determine the geographic position on earth. It has a 95% probability that the given position on the earth will be within 10-15 meters of the actual position. GPS allows precise mapping of the farms and together with appropriate software informs the farmer about the status of his crop and which part of the farm requires what input such as water or fertilizer and/or pesticides etc.

Geographic information system (GIS): It is software that imports, exports and processes spatially and temporally geographically distributed data.

Grid sampling: It is a method of breaking a field into grids of about 0.5-5 hectares. Sampling soil within the grids is useful to determine the appropriate rate of application of fertilizers. Several samples are taken from each grid, mixed and sent to the laboratory for analysis.

Variable rate technology (VRT): The existing field machinery with added Electronic Control Unit (ECU) and onboard GPS can fulfill the variable rate requirement of input. Spray booms, the Spinning disc applicator with ECU and GPS have been used effectively for patch spraying. During the creation of nutrient requirement map for VRT, profit maximizing fertilizer rate should be considered more rather than yield maximizing fertilizer rate.

Yield maps: Yield maps are produced by processing data from adapted combine harvester that is equipped with a GPS, i.e., integrated with a yield recording system. Yield mapping involves the recording of the grain flow through the combine harvester, while recording the actual location in the field at the same time.

Remote sensors: These are generally categories of aerial or satellite sensors. They can indicate variations in the colours of the field that corresponds to changes in soil type, crop development, field boundaries, roads, water, etc. Arial and satellite imagery can be processed to provide vegetative indices, which reflect the health of the plant.

Proximate sensors: These sensors can be used to measure soil parameters asuch as N status and soil pH) and crop properties as the sensor attached tractor passes over the field.

Computer hardware and software: In order to analyze the data collected by other Precision Agriculture technology components and to make it available in usable formats such as maps, graphs, charts or reports, computer support is essential along with specific software support.

Precision irrigation systems: Recent developments are being released for commercial use in sprinkler irrigation by controlling the irrigation machines motion with GPS based controllers. Wireless communication and sensor technologies are being developed to monitor soil and ambient conditions, along with operation parameters of the irrigation machines (i.e. flow and pressure) to achieve higher water use efficiency.

Precision farming on arable land: The use of PA techniques on arable land is the most widely used and most advanced amongst farmers. CTF (Contolled Traffic Farming) is a whole farm approach that aims at avoiding unnecessary crop damage and soil compaction by heavy machinery, reducing costs imposed by standard methods. Controlled traffic methods involve confining all field vehicles to the minimal area of permanent traffic lanes with the aid of decision support systems. Another important application of precision agriculture in arable land is to optimize the use of fertilizers especially, Nitrogen, Phosphorus and Potassium.

The global food system faces formidable challenges and that will increase over the next 40 years. More radical changes to the food system and investment in research are required to cope up with future challenges and their solutions. The decline in the total productivity, diminishing and degrading natural resources, stagnating farm incomes, lack of eco-regional approach, declining and fragmented land holdings, trade liberalization on agriculture, limited employment opportunities in non-farm sector, and global climatic variation have become major concerns in agricultural growth and development. Therefore, the use of newly emerged technology adoption is seen as one key to increase agriculture productivity in the future.

It is expected that application of balanced soft and hard PA technologies based on the need of specific socio-economic condition of a country will make PA suitable for developing countries also. 'Soft' PA depends mainly on visual observation of crop and soil and management decision based on experience and intuition, rather than on statistical and scientific analysis. 'Hard' PA utilizes all modern technologies such as GPS, RS, and VRT. Three components, namely, 'single PA technology', 'PA technology package' (for the user to select one or combination) and 'integrated PA technology', have been identified as a part of adoption strategies of PA in the developing countries.

Precision farming within the fruits & vegetables and viticulture sectors: In fruit and vegetable farming, the recent rapid adoption of automation systems for recording parameters related to product quality, allows growers to grade products and to monitor food quality and safety, including colour, size, shape, external defects, sugar content, acidity, and other internal qualities. Additionally, tracking of field operations such as spraying chemicals and use of fertilizers can be possible to provide complete fruit and vegetable processing methods.

Tools

Precision agriculture is usually done as a four-stage process to observe spatial variability. Precision agriculture uses many tools but here are some of the basics: Tractors, combines, sprayers, planters, diggers, which are all considered auto-guidance systems. The small devices on the equipment that uses GIS (geographic information system) are what makes precision ag what it is. You can think of the GIS system as the "brain." To be able to use precision agriculture the equipment needs to be wired with the right technology and data systems. More tools include Variable rate technology (VRT), Global positioning system and Geographical information system, Grid sampling, and remote sensors.

Data Collection

Geolocating a field enables the farmer to overlay information gathered from analysis of soils and residual nitrogen, and information on previous crops and soil resistivity. Geolocation is done in two ways:

- The field is delineated using an in-vehicle GPS receiver as the farmer drives a tractor around the field.

- The field is delineated on a basemap derived from aerial or satellite imagery. The base images must have the right level of resolution and geometric quality to ensure that geolocation is sufficiently accurate.

Variables

Intra and inter-field variability may result from a number of factors. These include climatic conditions (hail, drought, rain, etc.), soils (texture, depth, nitrogen levels), cropping practices (no-till farming), weeds and disease. Permanent indicators—chiefly soil indicators—provide farmers with information about the main environmental constants. Point indicators allow them to track a crop's status, i.e., to see whether diseases are developing, if the crop is suffering from water stress, nitrogen stress, or lodging, whether it has been damaged by ice and so on. This information may come from weather stations and other sensors (soil electrical resistivity, detection with the naked eye, satellite imagery, etc.). Soil resistivity measurements combined with soil analysis make it possible to measure moisture content. Soil resistivity is also a relatively simple and cheap measurement.

Strategies

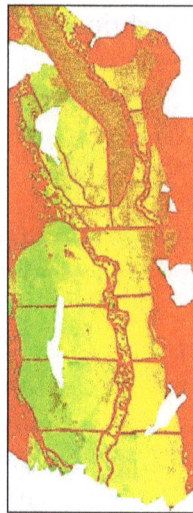

NDVI image taken with small aerial system Stardust II in one flight (299 images mosaic).

Using soil maps, farmers can pursue two strategies to adjust field inputs:

- Predictive approach: Based on analysis of static indicators (soil, resistivity, field history, etc.) during the crop cycle.

- Control approach: Information from static indicators is regularly updated during the crop cycle by:

  ○ Sampling: Weighing biomass, measuring leaf chlorophyll content, weighing fruit, etc.

  ○ Remote sensing: Measuring parameters like temperature (air/soil), humidity (air/soil/leaf), wind or stem diameter is possible thanks to Wireless Sensor Networks and Internet of things (IoT).

  ○ Proxy-detection: In-vehicle sensors measure leaf status; this requires the farmer to drive around the entire field.

  ○ Aerial or satellite remote sensing: Multispectral imagery is acquired and processed to derive maps of crop biophysical parameters, including indicators of disease. Airborne instruments are able to measure the amount of plant cover and to distinguish between crops and weeds.

Decisions may be based on decision-support models (crop simulation models and recommendation models) based on big data, but in the final analysis it is up to the farmer to decide in terms of business value and impacts on the environment – a role being takenover by artificial intelligence (AI) systems based on machine learning and artificial neural networks.

It is important to realize why PA technology is or is not adopted, "for PA technology adoption to occur the farmer has to perceive the technology as useful and easy to use. It might be insufficient to have positive outside data on the economic benefits of PA technology as perceptions of farmers have to reflect these economic considerations."

Implementing Practices

New information and communication technologies make field level crop management more operational and easier to achieve for farmers. Application of crop management decisions calls for agricultural equipment that supports variable-rate technology (VRT), for example varying seed density along with variable-rate application (VRA) of nitrogen and phytosanitary products.

Precision agriculture uses technology on agricultural equipment (e.g. tractors, sprayers, harvesters, etc.):

- Positioning system (e.g. GPS receivers that use satellite signals to precisely determine a position on the globe);

- Geographic information systems (GIS), i.e., software that makes sense of all the available data;

- Variable-rate farming equipment (seeder, spreader).

Usage around the World

The concept of precision agriculture first emerged in the United States in the early 1980s. In 1985, researchers at the University of Minnesota varied lime inputs in crop fields. It was also at this time that the practice of grid sampling appeared (applying a fixed grid of one sample per hectare). Towards the end of the 1980s, this technique was used to derive the first input recommendation maps for fertilizers and pH corrections. The use of yield sensors developed from new technologies, combined with the advent of GPS receivers, has been gaining ground ever since. Today, such systems cover several million hectares.

In the American Midwest (US), it is associated not with sustainable agriculture but with mainstream farmers who are trying to maximize profits by spending money only in areas that require fertilizer. This practice allows the farmer to vary the rate of fertilizer across the field according to the need identified by GPS guided Grid or Zone Sampling. Fertilizer that would have been spread in areas that don't need it can be placed in areas that do, thereby optimizing its use.

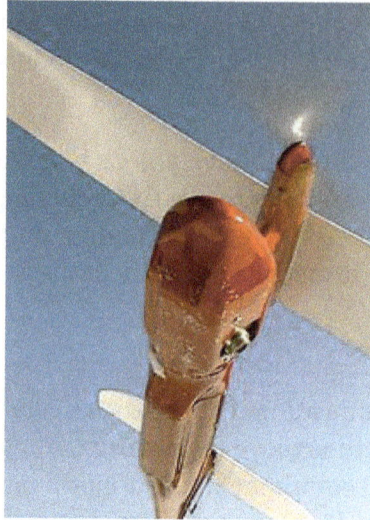

Pteryx UAV, a civilian UAV for aerial photography and photo mapping with roll-stabilised camera head.

Around the world, precision agriculture developed at a varying pace. Precursor nations were the United States, Canada and Australia. In Europe, the United Kingdom was the first to go down this path, followed closely by France, where it first appeared in 1997-1998. In Latin America the leading country is Argentina, where it was introduced in the middle 1990s with the support of the National Agricultural Technology Institute. Brazil established a state-owned enterprise, Embrapa, to research and develop sustainable agriculture. The development of GPS and variable-rate spreading techniques helped to anchor precision farming management practices. Today, less than 10% of France's farmers are equipped with variable-rate systems. Uptake of GPS is more widespread, but this hasn't stopped them using precision agriculture services, which supplies field-level recommendation maps.

One third of the global population still relies on agriculture for a living. Although more advanced precision farming technologies require large upfront investments, farmers in developing countries are benefitting from mobile technology. This service assists farmers with mobile payments and receipts to improve efficiencies. For example, 30,000 farmers in Tanzania use mobile phones for contracts, payments, loans, and business organization.

The economic and environmental benefits of precision agriculture have also been confirmed in China, but China is lagging behind countries such as Europe and the United States because the Chinese agricultural system is characterized by small-scale family-run farms, which makes the adoption rate of precision agriculture lower than other countries. Therefore, China is trying to better introduce precision agriculture technology into its own country and reduce some risks, paving the way for China's technology to develop precision agriculture in the future.

Economic and Environmental Impacts

Precision agriculture, as the name implies, means application of precise and correct amount of inputs like water, fertilizer, pesticides etc. at the correct time to the crop for increasing its productivity and maximizing its yields. Precision agriculture management practices can significantly reduce the amount of nutrient and other crop inputs used while boosting yields. Farmers thus obtain a return on their investment by saving on water, pesticide, and fertilizer costs.

The second, larger-scale benefit of targeting inputs concerns environmental impacts. Applying the right amount of chemicals in the right place and at the right time benefits crops, soils and groundwater, and thus the entire crop cycle. Consequently, precision agriculture has become a cornerstone of sustainable agriculture, since it respects crops, soils and farmers. Sustainable agriculture seeks to assure a continued supply of food within the ecological, economic and social limits required to sustain production in the long term.

Precision agriculture reduces the pressure on agriculture for the environment by increasing the efficiency of machinery and putting it into use. For example, the use of remote management devices such as GPS reduces fuel consumption for agriculture, while variable rate application of nutrients or pesticides can potentially reduce the use of these inputs, thereby saving costs and reducing harmful runoff into the waterways.

Emerging Technologies

Precision agriculture is an application of breakthrough digital farming technologies. Over $4.6 billion has been invested in agriculture tech companies—sometimes called agtech.

Robots

Self-steering tractors have existed for some time now, as John Deere equipment works like a plane on autopilot. The tractor does most of the work, with the farmer stepping in for emergencies. Technology is advancing towards driverless machinery programmed by GPS to spread fertilizer or plow land. Other innovations include a solar powered machine that identifies weeds and precisely kills them with a dose of herbicide or lasers. Agricultural robots, also known as AgBots, already exist, but advanced harvesting robots are being developed to identify ripe fruits, adjust to their shape and size, and carefully pluck them from branches.

Drones and Satellite Imagery

Advances in drone and satellite technology benefits precision farming because drones take high quality images, while satellites capture the bigger picture. Light aircraft pilots can combine aerial photography with data from satellite records to predict future yields based on the current level of field biomass. Aggregated images can create contour maps to track where water flows, determine variable-rate seeding, and create yield maps of areas that were more or less productive.

The Internet of Things

The Internet of things is the network of physical objects outfitted with electronics that enable data collection and aggregation. IoT comes into play with the development of sensors and farm-management software. For example, farmers can spectroscopically measure nitrogen, phosphorus, and potassium in liquid manure, which is notoriously inconsistent. They can then scan the ground to see where cows have already urinated and apply fertilizer to only the spots that need it. This cuts fertilizer use by up to 30%. Moisture sensors in the soil determine the best times to remotely water plants. The irrigation systems can be programmed to switch which side of tree trunk they water based on the plant's need and rainfall.

Innovations are not just limited to plants—they can be used for the welfare of animals. Cattle can be outfitted with internal sensors to keep track of stomach acidity and digestive problems. External sensors track movement patterns to determine the cow's health and fitness, sense physical injuries, and identify the optimal times for breeding. All this data from sensors can be aggregated and analyzed to detect trends and patterns.

As another example, monitoring technology can be used to make beekeeping more efficient. Honeybees are of significant economic value and provide a vital service to agriculture by pollinating a variety of crops. Monitoring of a honeybee colony's health via wireless temperature, humidity and CO_2 sensors helps to improve the productivity of bees, and to read early warnings in the data that might threaten the very survival of an entire hive.

Smartphone Applications

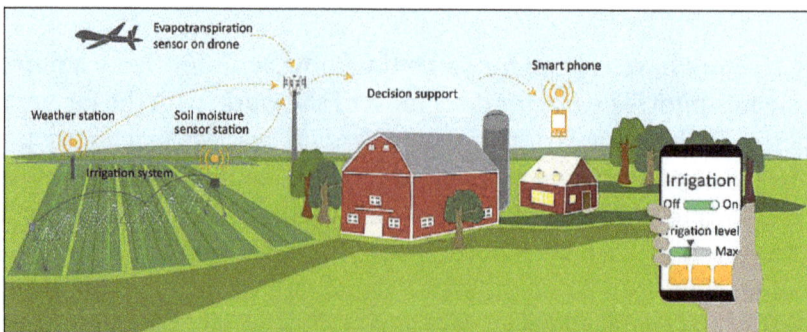

A possible configuration of a smartphone-integrated precision agriculture system.

Smartphone and tablet applications are becoming increasingly popular in precision agriculture. Smartphones come with many useful applications already installed, including the camera, microphone, GPS, and accelerometer. There are also applications made dedicated to various agriculture applications such as field mapping, tracking animals, obtaining weather and crop information, and more. They are easily portable, affordable, and have a high computing power.

Machine Learning

Machine learning is commonly used in conjunction with drones, robots, and internet of things devices. It allows for the input of data from each of these sources. The computer then processes this information and sends the appropriate actions back to these devices. This allows for robots to deliver the perfect amount of fertilizer or for IoT devices to provide the perfect quantity of water directly to the soil. The future of agriculture moves more toward a machine learning architecture every year. It has allowed for more efficient and precise farming with less human manpower.

Importance of Precision Agriculture

These days, farmers are increasingly under pressure to make critical decisions on the fly. Precision agriculture can help through innovative technologies that collect large amounts of information about fields and crops, analyze them, and deliver them to growers—all in real time.

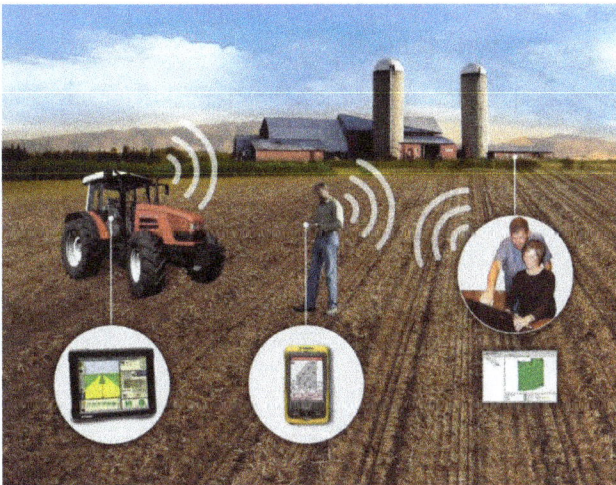

A robot sensor is one such device, which a farmer can install throughout his field to enable continuous management of his crops and harvest. The sensor essentially records and analyzes key data about the soil, crops, and water, and immediately sends the findings to the farmer.

Real-time data is also valuable in monitoring weather conditions, which can help farmers anticipate—and effectively overcome—issues like the adverse effects of extreme hot or cold weather.

Reduce Water Waste and Improve Crop Management

Drones and manned aircraft are also becoming widely used, particularly those with cameras mounted to them. For instance, when a drone flies over a field, the camera has the capability to capture dozens of high-resolution images in a matter of seconds. From there, it provides farmers with instant aerial views of the field that they would not normally see on the spot, informing them of any problem areas that need to be addressed immediately.

Similarly, aerial spectral imaging for precision agriculture is proving to be a useful "eye in the sky" that helps farmers manage crop production more efficiently throughout the entire year. This technology involves manned aircraft equipped with specialized camera systems that capture highly accurate images and data at specific wavelengths. The findings enable growers to employ biological and mathematical modeling to cross-reference the data to the actual physical state of the area, such as plant condition, water levels, and nitrogen and potassium content.

Based on the images he receives, a farmer can analyze soil fertility and hydration, identify insects and weeds, optimize the use of fertilizers and pesticides, and estimate crop yield.

Maximize Resources and Labor

Precision agriculture can also let farmers maximize their available resources without adding extra labor. One way is through a mapping tool that can let growers monitor field conditions with the goal of developing an optimal planting schedule. They can base their farming decisions on the findings received from the device, such as the ideal crop to plant at a certain time of the year; how much fertilizer to apply—and when—and which part of the field may require hydration.

Produce Food to Feed the Entire World

Perhaps the most significant benefit of precision agriculture is its potential to help farmers overcome the threat to global food security. Growers can implement the various technologies, in order to monitor and manage crop development. Other innovations include engineered lighting that develops technologically-enhanced vegetables; as well as biological pest control, which entails using living organisms to control and kill pests without the use of chemicals or pesticides.

It is safe to say that farmers now have more incentive to implement precision agriculture in their daily operations. With pressure now greater to deliver higher yield, an investment in innovative agricultural technology would be well worth it when it ultimately delivers increased profits, fosters a healthier environment and promotes sustainable farming.

Monitor Soil and Plant Parameters

Using digital agriculture tools, growers can determine peak conditions for plant growth and what nutrients their crops need to help them make their yield goals on a field by field basis. There are a number of technology options available today. One way is by placing sensors throughout the fields to record rainfall and soil conditions in these different areas. This can be helpful with VRT and improving yields in key problem areas. Often these sensors can be connected to digital or software applications to help growers maximize the benefits of this technology.

There are a variety of computer applications, software programs and other digital ag tools that can be used along with agricultural equipment or can be used manually to record soil sample results, fertilizer inputs, rainfall and other conditions. This technology helps agronomists and their growers evaluate their farming decisions for individual fields and improve their yields.

Automate Field Management

If a grower is using sensor technology in their operations, soil and plant species can be automatically optimized through these sensors and taken from a Decision Support System, which can help determine the best timing to water and fertilize the particular crops.

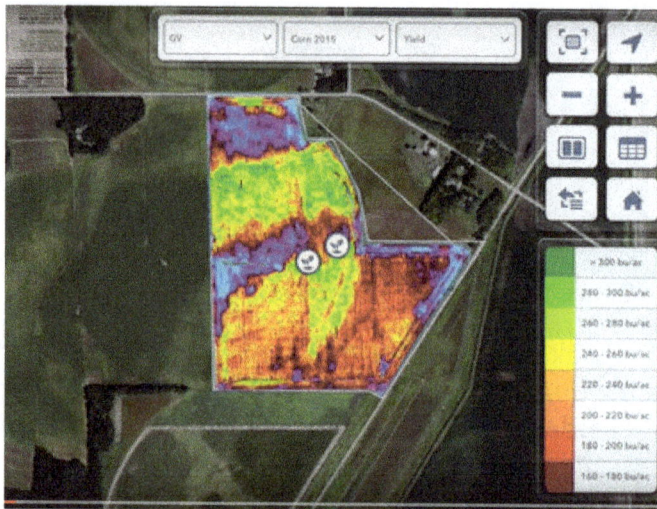

If a grower is using a digital application they can use a number of resources available through the platform to track progress during the season. One example includes field health imagery which allows growers to track the progress of the crop and support crop scouting efforts and field management decisions to help them maintain the highest yield potential as the season progresses.

Collect Real-time Data

If a grower applies sensing devices throughout the field, it will allow for continuous monitoring of the chosen parameters and offer real-time data to help the grower make informed decisions while planting, through the growing season and during harvest.

With digital ag tools, growers can use real-time data at key times to help them make decisions. For example, if a grower has uploaded their planting data or used any app during planting, they can build maps within their account. These maps can then be used during harvest or combined with harvest data to produce detailed yield results by field and by hybrid or variety.

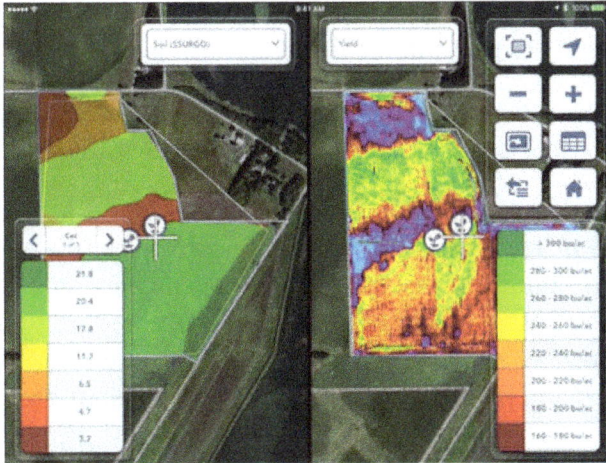

Get the Best Results from Labor and Resources

Growers and retailers are able to use technology to help maximize the benefits of crop nutrients, crop protection and irrigation costs by using automatic sensors that alert the grower of the need or best time to irrigate and fertilize their respective crops.

With digital ag tools, growers can create a variety of seeding and fertility prescriptions (scripts), including nitrogen, phosphorus, potassium and other essential nutrients.

It is estimated that more than half the growers currently engage in some form of agricultural technology and this number is expected to continue to grow as agronomists and growers see the improved benefits of digital ag tools and how they can improve their results within their overall precision agriculture strategies. For some digital applications there are even dealer options that allow the agronomist to track results and evaluate how crops performed across larger territories or among a variety of growing conditions.

Ag retailers have the opportunity to help their customers by providing them with local expertise on how the various digital ag tools and technologies available today might best be implemented within their geography and within their individual operation to help them improve their return on investment in the most appropriate ways.

Future Tasks to be Carried out with Precision Farming

Precision Plant Protection

The application of precision plant protection opened new perspectives in both the plant protection research and the plant protection practice. Precision plant protection can be seen as a GIS-based decision supporting system and way of farming, which considers the spatial heterogeneity of pest infestation at the arable site.

As it is well known the infestation of plant pathogens is very rarely homogeneous within a given area (field) their occurrence and spread is rather heterogeneous.

The aim of precision plant protection is to detect the varied damaging organisms accurately and to apply preventive technologies that can track this heterogeneous occurrence. Under extreme conditions we can find pathogens only on a very little part of the cultivated area, or the pathogen organisms are represented only under the threshold value, so we do not have to apply local treatments. With decisions like that we can save costs and reduce pesticide pressure on the environment (plant protection agents) considerably.

Precision plant protection includes three main activities:

- Space and time data and phenomenon acquisition on plant pathogens and on plant protection with high accuracy,

- GIS based data processing and analysis,

- High level automated site specific field work.

Whether the above three work processes are realized together on time and technical equipment or separately we can talk about on-line or real time realisation.

The main principle of the on-line method is that data acquisition is based on image-recording or detection. Further right after data analysis and processing we get the result, the process control command for the machine doing the treatment.

General scheme of a precision plant protecting treatment:

- Data acquisition,

- Forming a decision logistic system (algorithm),

- Selecting the system of equipment for application,

- Application of plant protecting agents.

Precision Weed Control

Applying precision weed control we consider the weed composition according to species and morpho-ecological groups on the investigated area. Investigations are carried out not only according to the aspects of chemical weed control, but also in all non-chemical applications (physical, mechanical etc.) belonging to the scope of integrated weed-control. It is almost indifferent whether process control is directed to a sprayer, mower or thermal equipment.

Generally we can say that it is reasonable to use precision methods in practice if there is a possibility to abandon local treatments due to the heterogeneity of weed population.

Weed survey should record the following criteria of weed species:

- Average cover,
- Frequency,
- Sequence of dominance,
- Morpho-ecological spectrum,
- Distribution of life-forms,
- Ratio of mono-and dicotyledons.

Steps of weed-control planning process:

- Marking the field contours by GPS,
- Making a sample-division plan,
- Visiting the sampling areas,
- Weed-cover survey,
- Evaluation of technological variants determining an ecological optimum,
- Programming the weed-killing machine process control,
- Carrying out the application,
- Post-checking, data-saving.

Precision Water Management

In most crops, growth can proceed unimpaired and crop yield can be maximized only when the soil moisture potential remains high (and water remains readily available) continuously throughout the growing season.

Water is critical to crop productivity since crop yields generally increase linearly with water transpired by a crop. Excess water (waterlogging) can induce nutrient and aeration stresses and encourage pests that reduce yield and quality. Water management is also critical to water quality because techniques to optimize water relations for plants can also impact fate and transport of pollutants to surface water and groundwater. Naturally, the adequacy of water for plant growth is primarily related to the amount, frequency and distribution of rainfall, soil properties as they affect processes that regulate soil water availability to plants and landscape properties that regulate the hydrologic cycle within a watershed. Three approaches to precision water management are therefore apparent:

- Variable rate irrigation,

- Matching agronomic inputs to water availability defined by soil and/or landscape properties,

- Drainage.

Variable Rate Irrigation

A well-managed irrigation system as one that optimizes the spatial and temporal distribution of water so as to promote crop growth and yield and to enhance the economic efficiency of crop production (maximum net return). He further states that since the physical circumstances and the socioeconomic conditions for irrigation are site specific (and often season specific) in each case, there can be no single solution to the problem of how best to develop and manage and irrigation management.

Considerable progress has been made with variable rate irrigation systems primarily with sprinkler irrigation provided by center-pivot and linear-move machines. These site-specific irrigation systems require high spatial resolution (currently 10-30 m) achieved by adding more discrete control between contiguous elements of the machine, all at higher costs that those of current systems. Variable irrigation is coupled with precision nutrient and pest management via chemigation, in part because variable irrigation facilitates in creased management precision in space and time and in part because it may not be economically feasible to site-specifically manage only for water. The uniformity of chemical application depends on the uniformity of water application, requiring injection equipment that can vary the amount of chemical injected into the boom in proportion to the fowl rate water in order to achieve the desired chemical application rate.

Success of precision irrigation management has been achieved with regard to application control. The key to the agronomic success of precision irrigation management depends to large extent on how well the water needs of the soil-plant system can be measured or predicted and the accuracy of water application prescriptions. The value of precision irrigation management depends on whether increased profits and the reduction in pollutants more than offset the cost of increased resolution needed in irrigation systems to apply irrigation and chemigate site specifically.

Due to the random variability in water application distributions due to wind, start-stop operations of the self propelled machines, and sprinkler pattern variations combined with the low cost of water and N fertilizers, it is probably not economically feasible to site-specifically manage only for water and/or nitrogen.

Soil-landscape Water Management

The potential for precision management of agronomic inputs increases with spatial variability in water availability within a field. Differences in water availability within a field are governed by:

- The occurrence of dissimilar soil types,

- The presence of soil degradation processes (*e.g.*, erosion, compaction and salinization),

- Variation in landscape.

The evidence for spatial variation in water availability is clear. North-facing slopes had 20% more available water in soils than south-facing slopes throughout the year. Where as soils on east-facing slopes were the driest. Crop yields are often highest in the lower slope positions where soil water and nutrient contents are higher. Eroded soils often have lower infiltration rates and lower available water than their non-eroded counterparts. Some portion of landscape variability can be attributed to the variation of soil properties with landscape position, whereas some is attributable to redistribution of water within a landscape due to either runoff or subsurface horizontal flow of water. Compacted soils reduce infiltration or restrict plant roots, thereby limiting water availability to plants. Areas of high salinity are known to reduce yields.

Knowledge of the spatial distribution of water availability can be used as a basis for site-specific input recommendations. There are three approaches for mapping soil water variability:

- County soil surveys,

- Interpolation of a network, usually a grid, of point samples to estimate spatial distribution of soil properties or water content,

- Soil-landscape models to estimate spatial patterns of soil water availability.

The presence of small scale spatial variability in soil physical properties and the high cost of network sampling may limit its use in mapping water availability. Site-specific soil water monitoring is used to some extent as a basis for variable rate irrigation and land in landscape studies. Statistical models of soil-landscape relationships offer opportunities to map spatial patterns of soil properties where relief or some landscape attribute is a primary factor contributing to soil variability. Soil-landscape models are important because terrain modifies the distribution of hydrologic and erosional processes (*i.e.*, soil water content, runoff and sedimentation) and soil temperature in fields, all important is regulating crop productivity of topography wit a regular grid of elevation observations is referred to as a digital terrain model (DTM) when attributes of a landscape are of interest and a digital elevation model (DEM) when merely relief is represented. A DTM allows the estimation of derivatives of elevation including slope, curvature, aspect, catchment area, and surface drainage proximity variables that correlate to soil and land qualities.

The value of DTM is that is increases the resolution of soil maps for use in site-specific management and in environmental modeling by using terrain attributes to spatially distribute estimated soil attribute data. Therefore, terrain modeling efforts have focused on its application to soil survey to model and depict the spatial variability of soil horizons in reference to the topographic surface and spatial application of simulation models to evaluate current and potential management practices regarding their effects on crop production and the environment in space and time. The extent of use of soil-landscape models is currently limited in field applications of precision agriculture. However, high-resolution DEMs can easily be created using DGPs and laser-based systems with high vertical accuracies. As elevation maps become available, soil-landscape modeling techniques such as DTM will be increasingly used in precision agriculture.

Drainage

Poor drainage is often cited by farmers as a source of yield variability within field. Many options for drainage currently exist and can be applied site-specifically. Therefore, there is little need to design site-specific drainage practices. The decision to install drainage is economically, not technically, limited, Regardless of scale, the decision to drain hinges on the expectation of returns on investment that exceed costs of installation. Site specifically, the cost of draining portions of fields or small isolated areas may be higher because the yield depression due to poor drainage can be accurately assessed if sufficient years are included in the calculation. Drainage, therefore, is a site-specific, economic decision based on the conditions at each site and cannot be generalized.

References

- Precision-agriculture-precision-farming: whatis.techtarget.com, Retrieved 03 February, 2019

- Remote-sensing-applications-in-agriculture, remote-sensing: grindgis.com, Retrieved 16 June, 2019

- Precision-agriculture: madhavuniversity.edu.in, Retrieved 08 January, 2019

- Major-advantages-of-precision-agriculture: mydecorative.com, Retrieved 02 February, 2019

- Benefits-precision-agriculture: inthefurrow.com, Retrieved 19 June, 2019

- Precision-water-management: agropedia.iitk.ac.in, Retrieved 05 March, 2019

Automatic Irrigation

Irrigation refers to the usage of controlled amounts of water to plants and crops at regular intervals. Automatic irrigation is the system used for the operation and management of irrigation structures. The diverse applications of automatic irrigation system have been thoroughly discussed in this chapter.

Automatic irrigation is the use of a device to operate irrigation structures so the change of flow of water from one bay, or set of bays, to another can occur in the absence of the irrigator.

SCADA automation on a pipe and riser outlet.

Automation can be used in a number of ways:

- To start and stop irrigation through supply channel outlets,

- To start and stop pumps,

- To cut off the flow of water from one irrigation area – either a bay or a section of channel and directing the water to another area.

These changes occur automatically without any direct manual effort, but the irrigator may need to spend time preparing the system at the start of the irrigation and maintaining the components so it works properly.

Benefits of Automatic Irrigation

Reduced Labour

As the irrigator is not required to constantly monitor the progress of an irrigation, the irrigator is available to perform other tasks – uninterrupted.

Improved Lifestyle

The irrigator is not required to constantly check the progress of water down the bays being irrigated. The irrigator is able to be away from the property, relax with the family and sleep through the night.

More timely irrigation: Irrigators with automation are more inclined to irrigate when the plants need water, not when it suits the irrigator.

Assists in the management of higher flow rates: Many irrigators are looking to increase the irrigation flow rates they receive through installing bigger channels and bay outlets. Such flow rates generally require an increase in labour as the time taken to irrigate a bay is reduced thus requiring more frequent change over. Automation allows for these higher flows to be managed without an increase in the amount of labour.

More accurate cut-off: Automation of the irrigation system allows cut-off of water at the appropriate point in the bay. This is usually more accurate than manual checking because mistakes can occur if the operator is too late or too early in making a change of water flow.

Reduced runoff of water and nutrients: Automation can help keep fertiliser on farm by effectively reducing run off from the property. Retaining fertiliser on farm has both economic and environmental benefits.

Reduced costs for vehicles used for irrigation: As the irrigator is not required to constantly check progress of an irrigation, motor bikes, four wheelers and other vehicles are used less. This reduces the running costs of these vehicles and they require less frequent replacement.

Disadvantages of Automatic Irrigation

Cost

There are costs in purchasing, installing and maintaining automatic equipment.

Reliability

Can the irrigator trust an automatic system to work correctly every time? Sometimes failures will occur. Often these failures are because of human error in setting and maintaining the systems. A re-use system is good insurance to collect any excess runoff when failures occur.

Increased Channel Maintenance

There is a need to increase maintenance of channels and equipment to ensure the system works correctly. Channels should be fenced to protect the automatic units from stock damage.

Pneumatic System

A pneumatic system is a permanent system activated by a bay sensor located at the cut-off point. When water enters the sensor, it pressurises the air, which is piped to a mechanism that activates the opening and closing of irrigation structures.

Portable Timer System

A portable timer system is a temporary system which uses electronic clocks to activate the opening and closing of the irrigation structures. Because of its portable nature, landowners usually buy 4 or 5 units to move around the whole property.

Timer/Sensor Hybrid

As the name suggests, this system is a hybrid of portable timer and sensor systems. Like a portable timer, it uses an electronic device to activate the opening and closing of the irrigation structures. However, this system has an additional feature of the irrigator being able to place a moveable sensor down the bay, which when comes in contact with water, transmits radio signals to the timer devices at the outlets to open or close the structures and sends a radio message to a receiver to let the landowner know water has reached the cut-off points down the bay.

SCADA

Automation systems that use Supervisory Control and Data Acquisition (SCADA) consist of a personal computer and software package to schedule and control irrigation via a radio link. Signals are sent from the computer to control modules in the paddock to open and close irrigation structures with linear actuators. Bays are opened and closed on a time basis, some systems have the capacity to automatically alter the time a bay outlet is open if the channel supply is inconsistent.

SCADA based systems have the additional benefit of being able to start and stop irrigation pumps and motors.

An irrigation layout can be automated at one of two places; in sections of channel or at individual bay outlets.

Automation of Channel Sections

In this system the channel structures are automated allowing the channel level to be changed. The bay outlets do not have opening or closing structures rather each set of outlets is set at a specific level eg a set of sills.

This method of automation requires a larger amount of fall to be available in the channel system to allow for a change in water level between different areas. This change in water level is required to prevent water flowing onto bays previously irrigated, when another section is to be irrigated. On many farms this fall is not available, so this method of automation in many cases is not suitable.

Automation of Individual Bay Outlets

This method of automation involves control of the bay outlets to change the flow of water onto the areas being irrigated. This system of automation is the most frequently used in areas where there is insufficient fall to automate channel sections.

The same type of automatic devices available can be set up to operate either automation of channel sections or automation of bay outlets.

All systems of automation have advantages and disadvantages that need to be considered when deciding which system will suit the irrigation layout for a particular property. There is no system that will be the "best" system for all properties.

The methods of irrigation used by the irrigator need to be considered. If a system that can be moved around the property and perhaps used on other properties is required, then the irrigator needs to consider those systems that are portable. If the irrigator wants a system where the components are fixed and can follow the same irrigation sequence each irrigation, then a fixed system would be more appropriate.

In determining the best system for a property, the irrigator will need to consider the cost of the system, back up servicing of the system and which system will best suit the property and irrigation layout.

SCADA Actuator Attached to a Farm Channel Bay Outlet

Development of a whole farm plan for the property is a good way to start preparing for automation. During the development of a whole farm plan, landholders should consider automatic irrigation in the planning process so they can incorporate some of the features required for automation from the start. This might involve design of the channels for channel automation if that is possible or it might be the use of bay outlets and other channel structures that will suit automation at a later stage.

When it comes to starting to install automation there are a number of ways of getting started. One way is to start by automating those areas irrigated at night, so appropriate irrigation flow rates can be acheived, without disrupting the irrigator's sleep. Another is to automate those areas that are difficult to irrigate – areas of short steep bays that require the irrigator to be present more often or require frequent changes.

Automation is not only suited to areas of the farm that have been laser graded. Non-lasered areas can also be automated. This can include automation of the channel structures to irrigate sections of the non-lasered areas.

With the information from a whole farm plan, channel structures that will be used when the development works are carried out, can be purchased and used to automate these non-lasered areas. This can be done with the knowledge the structures will be suitable for use after the development work is carried out.

Automatic Plant Irrigation System

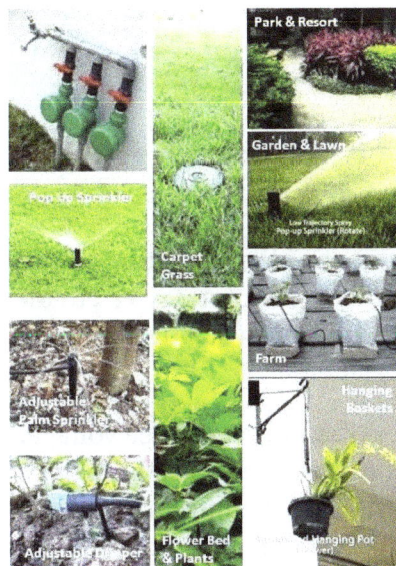

Automatic Plant Irrigation System.

In present days, in the field of agriculture farmers are facing major problems in watering their crops. It's because they don't have proper idea about the availability of the power. Even if it is available, they need to pump water and wait until the field is properly watered, which compels them to stop doing other activities – which are also important for them, and thus they loss their precious time and efforts. But, there is a solution – an automatic plant irrigation system not only helps farmers but also others for watering their gardens as well.

This automatic irrigation system senses the moisture content of the soil and automatically switches the pump when the power is on. A proper usage of irrigation system is very important because the main reason is the shortage of land reserved water due to lack of rain, unplanned use of water as a result large amounts of water goes waste. For this reason, we use this automatic plant watering system, and this system is very useful in all climatic conditions.

The power supply consists of a step-down transformer, which steps down the voltage to 12VAC. By using a bridge rectifier this AC is converted to DC, then it is regulated to 5v using a voltage regulator which is used for the operation of the microcontroller.

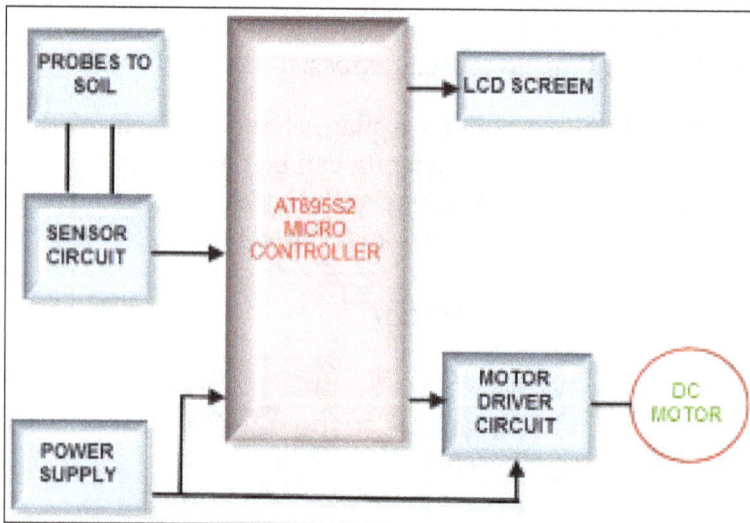

Block Diagram of Automatic Plant Irrigation System.

The block diagram of this automatic plant irrigation system comprises three main components namely a microcontroller, a motor-driver circuit and a sensor circuit. When the sensor circuit senses the condition of soil, it compares it with the reference voltage 5v. This process is done by a 555 timer.

When the soil condition is less than the reference voltage, i.e., 5v, then the soil is considered as dry and instantly the 555 timer sends the logic signal 1 to the microcontroller. The microcontroller then turns on the motor driver circuit and prompts the motor to pump water to the plants. When the soil condition is greater than the reference voltage, the soil becomes dry. Then the timer sends the logic signal 0 to the microcontroller, this turns off the motor driver circuit and prompts motor to pump water to the fields. Finally, the condition of the motor and soil are displayed in the LCD display.

Circuit Diagram of Plant Irrigation System

The main component used in this automatic plant irrigation system is 7404 Hex Inverter. The main function of the inverter output is proportional to input. It means, if the input of the inverter is low, then the output of the inverter will be high, and the inverter will give low output if the input is high. The Hex inverter 7404 IC includes six independent inverters and the range of operating voltage is around 4.75 V to 5.5 V, and the Supply voltage is 5 V. They are used in many applications such as drivers, inverting buffers, etc. This IC is available in different packages like quad-flat package and dual-inline package. The pin configuration of the 7404 IC is shown below.

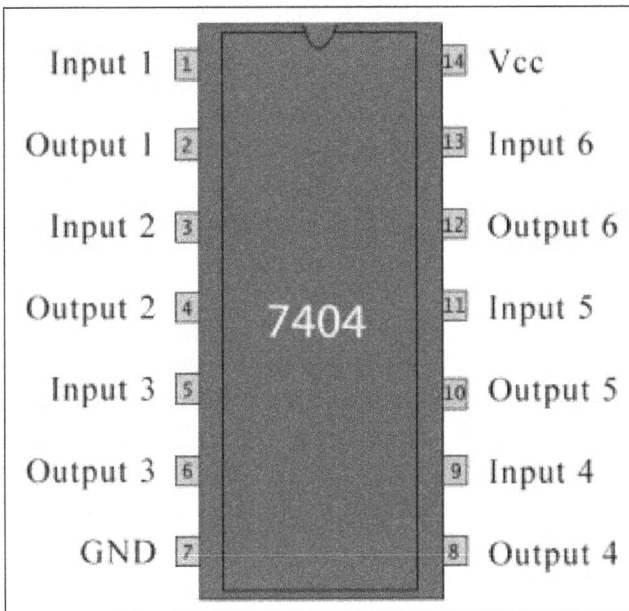

7404 IC Pin Configuration.

The circuit diagram of the plant-irrigation system is shown below. To make the circuit work and to water the pants, we use this simple logic: when the soil is dry, it has high resistance and when the soil is wet it has low resistance. This circuit consists of two probes that are placed into the earth. These probes perform the work only when the soil resistance is low and they cannot perform when the resistance of the soil is high.

To conduct the probes, the voltage supply is provided from a battery, which is connected to the circuit. When the soil becomes dry, it produces large voltage drop due to high resistance, and this is sensed by the hex inverter and makes the first NE555 timer. This timer is arranged as a monostable multivibrator with the help of an electrical signal.

When the first 555 timer is activated at pin2, it generates the output at pin3; and, this output is given to the input of the second timer. This second NE555 timer is configured with astable multivibrator and generates the output to make the relay which is connected to the electrically operated value through the SK100 transistor. The output of the second timer switches on the transistor that drives the relay. This relay is connected to the input of an electrical value and the output of the electrical value is given to the plants through the pipe.

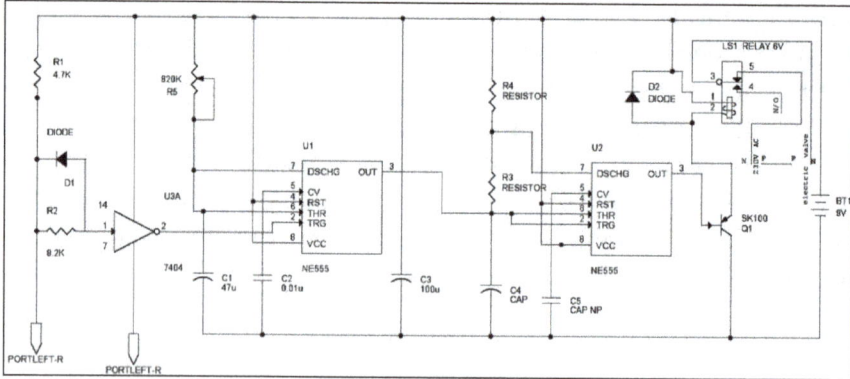

Plant Irrigation System Circuit Diagram.

When the relay is turned on, the valve opens and water through the pipes rushes to the crops. When the water content in the soil increases, the soil resistance gets decreases and the transmission of the probes gets starts to make the inverter stop the triggering of the first timer. Finally the valve which is connected to the relay is stopped.

The advantage of using an automatic plant irrigator is that it is a simple system capable of conserving water, improving growth, discouraging weeds, saving time, and controlling fungal diseases and adaptable to the conditions.

Micro Irrigation System

Micro Irrigation System.

Micro irrigation is nothing but a slow and regular application of water and nutrients moving down drop-by-drop directly to the root zone of the plants through low-discharge emitters and plastic pipes. This irrigation system is today's need of the hour as the natural water resources which are gift to the mankind have become scarce, and that are now not unlimited and free forever. But, the world's water resources are now fast moving back on track. After one completes the study of inter relationship between crops, soil, water and climatic conditions, one will find this micro irrigation system as a suitable system capable of delivering exact quantity of water at the root zone of the plants.

This system ensures that the plants do not endure from the strain or stress of less and over watering. The advantages of using this micro irrigation system are that for every drop of water used, we get more crop, better quality, early maturity, higher yield. Moreover, this system saves labor cost and water up to 70%. The working of this irrigation system covers over 40 crops spanning across 500 acres.

Automatic Plan Irrigation System using Microcontroller

Automatic Irrigation System on Sensing Soil Moisture Content

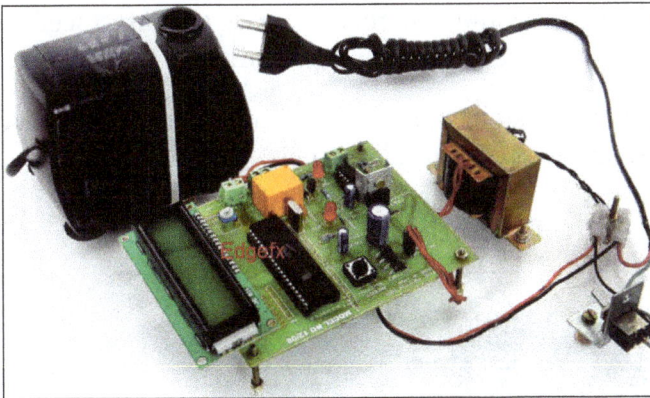

The automatic irrigation system on sensing soil moisture project is intended for the development of an irrigation system that switches submersible pumps on or off by using relays to perform this action on sensing the moisture content of the soil. The main advantage of using this irrigation system is to reduce human interference and ensure proper irrigation.

The Microcontroller acts as a major block of the entire project, and a power supply block is used for supplying power of 5 V to the whole circuit with the help of a transformer, a bridge rectifier circuit and a voltage regulator. The 8051 microcontroller is programmed in such a way that it receives the input signal from the sensing material which consists of a comparator to know the varying conditions of the moisture in the soil. The OP-AMP which is used as comparator acts as an interface between the sensing material and the microcontroller for transferring the moisture conditions of the soil, viz. wetness, dryness, etc.

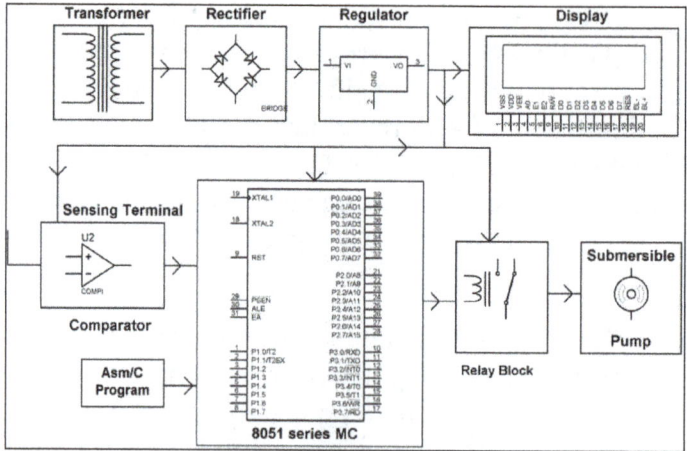

Block Diagram of Soil Moisture Content Based Irrigation.

Once the microcontroller gets the data from the sensing material – it compares the data as programmed in a way, which generates output signals and activates the relays for operating the submersible pump. The sensing arrangement is done with the help of two stiff metallic rods that are inserted into the agricultural field at some distance. The required connections from these metallic rods are interfaced to the control unit for controlling the operations of the pump according to the soil moisture content.

This automatic irrigation system can be further enhanced by using advanced technology that consumes solar energy from solar panels.

Solar Powered Auto Irrigation System

Solar Powered Auto Irrigation System Circuit.

The power from utilities is required to operate the system, this system uses solar panels to power the circuit. In agricultural field, the proper usage of automatic irrigation method is very vital due to some shortcomings of the real world like scarcity of land reservoir water and scarcity of rainfall. The water level (the ground water table) is getting reduced due to continuous extraction of water from the ground and thus gradually resulting in water scarcity in the agricultural zones slowly turning them into barren lands.

In the above irrigation system, solar energy generated from the solar panels is used for operating the irrigation pump. The circuit comprises moisture sensors built by using OP-AMP IC. The OP-AMP is used as comparators. Two stiff copper wires are inserted into the soil to know whether soil is wet or dry. A charge controller circuit is used to charge the photovoltaic cells for supplying the solar energy to the whole circuit.

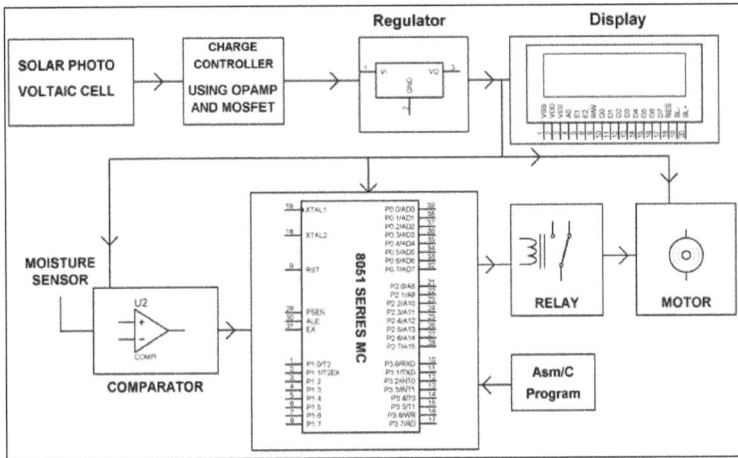

Solar Powered Auto Irrigation System Blockdiagram.

A moisture sensor is used for sensing the soil condition – to know whether the soil is wet or dry, and the input signals are then sent to the 8051 microcontroller, which controls the whole circuit. The microcontroller is programmed by using KEIL software. Whenever the soil condition is 'dry', the microcontroller sends commands to the relay driver and the motor gets switched on and supplies water to the field. And, if the soil gets wet, the motor gets switched off.

The signals that are sent from the sensors to the microcontroller through the output of the comparator operate under the control of a software program which is stored in the ROM of the microcontroller. The LCD displays the condition of the pump (on or off) interfaced to the microcontroller.

This automatic irrigation system can be further enhanced by using GSM technology to gain control over the switching operation of the motor.

GSM Based Automatic Irrigation System

Nowadays farmers are struggling hard in the agricultural fields round the clock. They do their field work in the morning section and irrigate their land during night time with intermittent intervals. The task of irrigating fields is becoming quite difficult for the farmers due to lack of regularity in their work and negligence on their part because sometimes they switch on the motor and then forget to switch off, which may lead to wastage of water. Similarly, they even forget to switch on the irrigation system, which again leads to damage to the crops. To overcome this problem, we have implemented a new technique by using GSM technology, which is explained below.

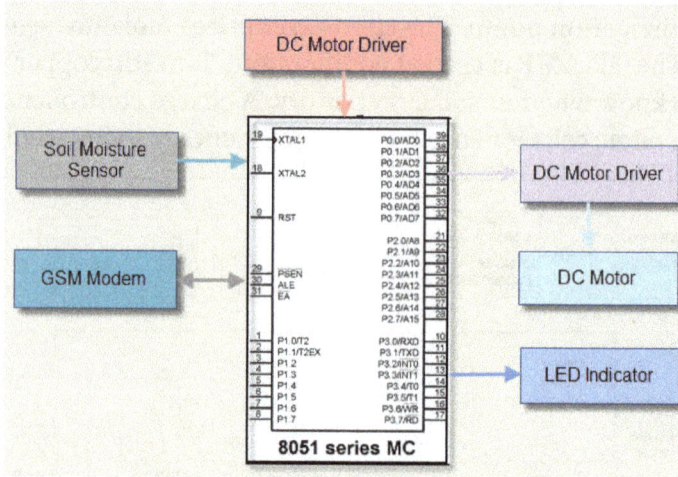

GSM Based Automatic Irrigation System.

The GSM Based automatic irrigation system is a project in which we get update status of the operation carried out in the agricultural fields via SMS with the help of a GSM modem. We can also add other systems such as LCD displays, web cam and other smart controlled devices.

We are using soil moisture sensor which is used to sense the moisture level in the – to know whether it is dry or wet. The moisture sensor is interfaced with the microcontroller. The input data signals from the moisture sensor are sent to the microcontroller and based on that it activates the DC Motor and switches the motor on with the help of a motor driver. After the soil gets wet, the Motor gets switched off automatically. The status of the agricultural fields can be known from the indication of the Light Emitting Diode (LED) or through the message sent to the GSM modem placed at the field. Simultaneously it is possible to send messages through a mobile to kit through the GSM modem. Thus, the irrigation motor can be controlled by using a mobile and a GSM modem.

Automated Irrigation Systems

Hand watering is the most common method of watering plants because it's fairly easy and can be done for little more than the cost of the water itself. Unfortunately, because about half of all household water goes to the yard or garden, hand watering is also the quickest way to waste water—not to mention time. You have to run back and forth to fill up the watering can, or if you're using a hose, you're likely giving the pavement an unnecessary drink.

If you're hand watering, chances are you're either under- or over-watering your plants as well. Many plants fare better when watered in small amounts several times a day—and how many of us really hand water anything that often? Plants usually cannot absorb all the water and fertilizer that gets dumped on top of them at once. There's also no faster way to spread bacterial and fungal diseases to all your plants than by watering them in this manner. If even a single leaf or stem harbors disease-causing fungi or bacteria, water can pick those organisms up, and as the infected water drips onto the leaves of other plants and into the soil, the bacteria and fungi are transplanted there as well—thus infecting the rest of your yard or garden. Keep in mind that this is a danger with any overhead watering technique.

How is Automatic Irrigation Better?

Automated irrigation systems can save homeowners the trouble of even having to think about watering their lawns, although checking up on each plant occasionally to make sure the sprinkler system is doing its job is highly recommended. Because automated irrigation systems are usually intended to fertilize plants as well as water them, they can also cut back the amount of time spent fertilizing by hand. Installers will tailor your sprinkler system layout to the various species of plants throughout your yard or garden to ensure they all receive the appropriate amount of water and nutrients at the right frequency.

Automated irrigation or sprinkler systems are also a fantastic way to save water, particularly if you have a root zone system installed. Because plants take in water through their roots, the closer to their roots the water originates, the more water they are able to absorb before it evaporates or just travels away. Systems can even be installed in such a way as to avoid runoff, which is a major water waster.

It is possible to get all sorts of bells and whistles on your automatic irrigation system, too—things like soil or rain sensors that will let your system controller know when it has been or is supposed to be raining a lot, or if the soil around your plants is already adequately damp. If you decide to get a system without such sensors, make sure you manually switch your controller to account for weather events like heavy rainfall and increased or decreased temperatures. At the very least, you should alter the settings on your controller every season; plants' water needs change the most drastically when the seasons shift.

Automated irrigation systems save time and money as long as they are programmed to make the most of the conditions in your yard or garden.

The Benefits of Automated Irrigation Systems

Many home owners are under the wrong impression that using the hose "like their father did" to water the lawn is more cost efficient than installing an automated irrigation system. On the contrary, installing automated irrigation systems in CT is an effective way to reduce your water costs, prevent uneven watering and keep your lawn healthier. Additionally, installing an efficient irrigation system can also increase the value of your property and save you time. These are the top 3 problems solved by installing an automated irrigation system:

Prevents Uneven Watering

When we install irrigation systems, we spend a good amount of time planning the layout so that the radius of each nozzle distributes the water where it needs to go and does so evenly, preventing over or under-watering. While designing such a layout, we take into account any slopes, along with other factors, which may cause water to flow and not get properly absorbed into the soil.

While it is possible to achieve even watering with a hose, it is time consuming and not as reliable as a programmed sprinkler.

Healthy Lawn

Like most ailments, prevention is much less costly than the cure. If you water and care for your lawn properly, you are much less likely to have to spend on replacing it or to bring it back to life. When using a hose to water your lawn, you risk the chance of uneven watering as described above. Besides causing your water to be used inefficiently, it also can case the soil to not retain nutrients, and therefore wither. A professionally-installed sprinkler system can help ensure that your soil and thus your lawn continue to be healthy.

Additional Benefits

In addition to reducing the cost of your water bills and keeping your lawn healthy, the right automated irrigation system can help increase the value of your home, as well as save you time. Most new home buyers today are expecting a home with amenities like the irrigation system. They are not looking to spend 2 – 5 hours a week watering their home with a hose, and that is most likely the time you would like to get back into your life as well.

One of the most obvious advantages is the time savings afforded by an automatic sprinkler or drip irrigation system. Once installed, many systems can be set to a timer to water at specific time intervals and on certain days of the week. This means there's no need to worry about forgetting to water the lawn and coming back from vacation to find crisp, yellow grass.

Another advantage is that irrigation systems, particularly the drip type, can be positioned so that water is more effectively targeted where it is needed. Nozzles can be adjusted, and underground drip tubes will deliver water right to the roots, rather than spraying walkways and driveways.

Another advantage is that automatic irrigation systems are generally hidden from view, which means there are no unsightly hoses stretched across the lawn and no more tripping hazards. Sprinkler heads pop up to spray and then retract when the job is done. Underground drip systems do their work out of view. For families with young children and pets who share outdoor spaces, automatic systems may be a safer option.

One of the most obvious advantages is the time savings afforded by an automatic sprinkler or drip irrigation system. Once installed, many systems can be set to a timer to water at specific time intervals and on certain days of the week. This means there's no need to worry about forgetting to water the lawn and coming back from vacation to find crisp, yellow grass.

Another advantage is that irrigation systems, particularly the drip type, can be positioned so that water is more effectively targeted where it is needed. Nozzles can be adjusted, and underground drip tubes will deliver water right to the roots, rather than spraying walkways and driveways.

Another advantage is that automatic irrigation systems are generally hidden from view, which means there are no unsightly hoses stretched across the lawn and no more tripping hazards. Sprinkler heads pop up to spray and then retract when the job is done. Underground drip systems do their work out of view. For families with young children and pets who share outdoor spaces, automatic systems may be a safer option.

References

- Automatic-irrigation, farm-management-soil-and-water-irrigation: agriculture.vic.gov.au, Retrieved 18 August, 2019

- Automatic-plant-irrigation-system-circuit-and-its-working: watelectronics.com, Retrieved 18 June, 2019

- Microcontroller-based-automatic-irrigation-system: elprocus.com, Retrieved 17 January, 2019

- Spread-the-love-to-your-plants-by-automating-your-irrigation: bestpickreports.com, Retrieved 09 June, 2019

- The-benefits-of-automated-irrigation-systems: summerrainsprinklers.com, Retrieved 25 August, 2019

- Pros-cons-of-automatic-irrigation-systems: tulsaworld.com, Retrieved 23 July, 2019

Harvest and Post-harvest Automation

Some of the equipment that are used in harvest automation are grain cart, conveyor belt, haulm toppers, etc. Dehulling system and sorting system fall under the domain of post-harvest automation. The topics elaborated in this chapter will help in gaining a better perspective about harvest and post-harvest automation.

Harvest Automation

When looking at modern farm equipment, specifically equipment utilized to produce grain crops, the trend has been to higher power machines. For example, today it is common to see 450 kW tractors on farms. To effectively utilize the power produced from the engine, the tractor must be adequately ballasted. In general there is a recommendations the tractor be ballasted at 60 to 70 kg per kW of engine power, or from 27,000 to 31,500 kg (60,000 to 70,000 lb_f) total mass. Of course when ballasting a tractor it is not permissible to exceed tire manufacturer's recommendations for load and inflation pressures. In fact, because of the soil-tire interface, common practice dictates that tire inflation pressures be reduced to the absolute minimum to achieve the best possible performance and fuel efficiency. As tractor size increase above the current upper limits, one or more of the following limitations must be overcome: 1) allowable tire loads must increase for limited section sizes; 2) tires must be added to axles (i.e, duals and triples); 3) tire diameters must increase; or 4) drive trains must reconfigured to include more than two axle. The dilemma in European is that tractor manufacturers must work within the 3.0 and 3.5 m transport widths thereby limiting tire spacing and/or section widths. By today's standards it is impractical to achieve axle loads in excess of 15,000 kg (33,000 lb_f). The two viable options that remain are larger diameter tires, or more axles.

When matching tillage tools and seeding equipment with available power, it is common to see fully loaded no-till planter develop draft forces approaching 2,000 N/drill row (450 lb_f/drill row) from ASABE (2009). Assuming a seeding speed of 10.0 km/h (6.0 mi/h), this implement requires tractor engine power approaching 9.0 kW/row (12.0 hp/row). Putting this in perspective, a 36 row no-till planter will require 325 kW (430 hp) tractor assuming a tractive efficiency of 77% and a transmission efficiency of 90% for a four wheel drive (4WD) tractor. It is the combination of implement width, ground speed, draft and tractive efficiency that mandate the minimum tractor size. The tractor must be ballasted to take full advantage of the engine power. Typically, ballasted tractor mass be range from 64 to 67 kg per engine kW (105 to 110 lb_f/Hp) for a total tractor mass of around 21,000 kg (46,000 lb_f). With a 60/40 static weight split between the front and rear axles, as is typical of properly ballasted 4WD tractors, and assuming row-crop dual tires, each tire must support a load of up to 3,150 kg/tire (6,950 lb_f/tire). From manufacturer specifications the minimal acceptable tire is 480/80R42 at an inflated pressure of 48 kPa. When going to single tires the minimal acceptable tire size is a 900/50R42, again inflated to 48 kPa. For row crop tires the minimal tractor width is 3.53 m (11.57 ft) while for single tires the minimum width is 2.84 m (9.32 ft). The latter case is what most European producers are required to accept.

Increasing width quandary - Many agricultural producers utilize large equipment to reduce labor costs and improve timeliness of their operations. In terms of spray application, producers have turned to faster sprayers with boom widths in excess of 30 m. Pesticide application errors, especially those associated with larger equipment, result in costly over application and reduced yield from crop injury or poor pest control. Over-application tends to increase with boom section width as operators attempt to control boom sections manually. A recent study found that manual operation of a 24.8 m boom (5 control sections) resulted in an average over-application of 12.4% across a wide range of field shapes and sizes.

Summary Statistics for Modern Field Machinery Power and Mass.			
Equipment Make and Model	Unballasted Mass (kg)	Ballasted/Loade Mass (kg)	Engine Power (kW)
AGCO MT975B 4WD Tractor	22,900	27,200	464
Case IH Axialflow 9120 Combine w/ 16 Row Corn Head	21,500	31,600	390
Balzer 2000 Grain Cart (54.5 T Capacity)	14,800	69,300	-
AGCO Rogator 1396 SP Sprayer (4,160 L Tank)	13,700	17,860	323

Off-rate application errors also result from the velocity differential across the spray boom that occur when spraying while turning, pressure variation across the spray boom, and undulating terrain which affects boom-canopy distance causing irregularities in nozzle pattern overlap. Previous research has indicated that off-rate errors resulting from turning movements on a sprayer with a 24.8 m boom could affect between 3% and 23% of fields (variety of shapes and sizes) receiving an application rate beyond ±10% of the target rate (Luck et al., 2010b). Problems associated with off-rate application errors are exacerbated with larger equipment as increased boom widths result in greater velocity, pressure, and height variations across the spray boom.

Evolving Automation Technologies

Looking towards the future to a point in time when humans are removed from field machinery, there are several emerging technologies that will be essential for autonomous operation. In some cases infrastructure development such as densification of Real-Time Kinematic (RTK) GPS networks to generate Virtual Reference Stations (VRS) correction data along with the development of Internet connectivity via Wi-Fi and WLAN to support data transfer. What follows is a brief overview of the status of many of the allied technologies that will be essential for totally autonomous field machinery of the future.

Space-based positioning systems - Advancements in sensing, communication and control technologies coupled with Global Navigation Satellite Systems (GNSS) and Geographical Information Systems (GIS) are aiding the progression of agricultural machines from the simple, mechanical machines of yesterday to the intelligent, autonomous vehicles of the future.

The U.S. Global Positioning System (GPS) is maintained by the U.S. government and has been in operation since the late 1970s. The benefits of GPS, specifically in the agricultural industry, have been well documented as they have progressed from point location mapping (soil sampling or yield monitoring) to real-time equipment control (auto-steer or map-based automatic section control). To increase the accuracy of the existing GPS network, additional technologies have been developed by both public and private institutions. The Nationwide Differential GPS System (NDGPS) was developed for use in the U.S. and included beacons maintained by the U.S. Coast Guard and the Department of Transportation. The Wide Area Augmentation System (WAAS) is operated by the Federal Aviation Administration. The WAAS network has become available for a variety of other users desiring sub-meter accuracy who have compatible receivers. A more recently developed system for improving GPS accuracy is the Continuously Operating Reference Stations (CORS) that was initially created by the National Oceanic and Atmospheric Administration. Since its inception, additional organizations have joined the network and provide correction data from their land-based GPS stations.

The Global Navigation Satellite System (GLONASS) is a Russian-operated satellite network that was developed in the late 1970s and was extended to non-military use in 2007. GLONASS is comparable to the U.S. GPS system and was created to provide real-time positioning data to compatible receivers. The GLONASS system is continually upgraded as existing satellites exceed their service life and new series replace them. The GLONASS-M series is currently in operation, with the GLONASS-K1 series expected to become operational in 2011.

The Galileo global navigation satellite system is currently being developed by the European Union (EU) to provide a separate network of satellites from the Russian and U.S. systems that are now in use. The Galileo system has been developed by the European Space Agency primarily to provide real-time positioning data for civilian use and was designed to be compatible with the Russian and U.S. systems. Two experimental satellites have been successfully launched and four additional satellites are planned to be launched in 2011 to validate system operation.

The accuracy of differential global position systems (DGPS) degrade with increasing distance to the reference station. For DGPS systems, an inter-receiver distance of a few hundred kilometers will yield a sub-meter level accuracy, whereas for Real Time Kinetic (RTK) systems a centimeter level accuracy is obtained for distances of less than 10 km. To service larger areas without compromising on the accuracy, several reference stations have to be deployed. Instead of increasing the number of real reference stations, Virtual Reference Stations (VRS) are created from the observations of the closest reference stations. The locations of the VRS can be selected freely but should not exceed a few kilometers from the rover stations. Typically one VRS is computed for a local area and working day.

The observations from the real reference stations are used to generate models of the distance dependent biases. Individual corrections for the network of VRS are predicted from the model parameters and the user's position. This kind of network applied to DGPS and RTK systems is known as wide-area DGPS (WADGPS) and network RTK respectively. An example of a commercially available network RTK is Trimble's VRS that provides high-accuracy RTK positioning for wider areas. A typical VRS network set up consists of GNSS hardware, communications interfacing and, modeling and networking software. Most of the existing network RTK systems have been installed in the densely populated areas of central Europe.

Wireless communications - For large scale high-tech agricultural operations, establishing vehicle to vehicle and vehicle to office communication is becoming imperative to manage the logistics of the tasks and to ensure the safety of the machines working in the field. The capability to transfer data wirelessly can help monitor the working statuses of these machines and allow dynamic reallocation of tasks in the event of malfunctions. Point to point and point to multi point communication can specifically be used for leader-follower systems. Cell GSM, Wi-Fi, WLAN and Wireless stand-alone modems are typically used for vehicle to vehicle and vehicle to office communications. These technologies compete with each other with regards to bit rate, mobility of terminals, signal quality, coverage area, cost and the power requirements. WLANs are used for high bit rate transfers whereas cellular GSM networks are used for large coverage areas. From a cost and power requirement perspective, cellular networks are far more expensive to establish and maintain than WLAN access points. The power requirement for a cell phone to transmit can be as high as several hundred milliwatts, while WLAN requires a maximum of 100 milliwatts (Wireless Center, 2010). In terms of mobility and controlled signal quality cellular GSM are superior to WLANs. WLANs suffer from low mobility, isolated coverage and vulnerability to interference. Each technology is strong where the other is weak and hence WLAN and cell GSM networks are complementary.

WLANs operate in the 2.4GHz unlicensed frequency band. The signaling rate is 11Mbps, and the terminals employ CSMA/CA (Carrier Sense Multiple Access with Collision Avoidance) to share the available radio spectrum. The distance between the transmitter and the receiver has the greatest influence on the signal quality and the thus the quality worsens with increase in the distance. For a 2.4 GHz spectrum band, if the distance is within 28 meters the data transfer rate can be up to 11Mbps whereas, for distances greater than 55 meters the transmission cannot be more than 1 Mbps. A GSM signal occupies a band width of 200 KHz and can have channel rates of up to 271 Kbps. The strengths of both cell GSM and WLANs are provided by wireless internet (Wi-Fi). These networks provide a coverage range of up to 600 ft (183 m) and operate typically at a frequency of 2.4GHz.

On-vehicle communications - With the introduction of microcontrollers to agricultural filed machinery it was not long until equipment designers realized the need to share and manage information between controllers. Following the lead of the truck, bus and automotive industries, equipment designers began looking for bus configurations and data structures to support continuing machinery development. Quickly, most designers realized the need for standardization to facilitate interoperability and interchangeability the industry came to grips with for hitching (ISO 730, 2009) and hydraulic systems. The following discussion highlights some of the more significant milestones in the evolution of the of on-vehicle communications and concludes with a brief treatment of what the industry can expect in the near future.

The Landwirtschaftliches BUS-System (LBS) is regarded as the precursor to ISOBus. Development of this protocol began in Germany in the late 1980s by a committee formed from the German Farm Machinery and Tractor Association. CAN version 1 was used as the base for developing this new agricultural communication bus protocol. The protocol was developed with the goal of running distributed process control systems such as fertilizer distribution, pesticide application, and irrigation. Therefore, development on the protocol began with the goal of standardizing network data exchange between electronic components on agricultural tractors and implements.

Based on the preliminary work by Auernhammer in Germany, ISO was requested to begin development of a standardized protocol for agricultural equipment in the early 1990s.

ISOBus is a distributed network protocol specification (developed under ISO 11783) for equipment which utilize CAN technology for electronic communication in the agricultural industry. Development of this ISO protocol began in the early 1990s when a working group was formed to develop an interim connector standard (ISO 11786). In 1992, ISO 17783 was formed to continue development of the communications protocol standard. Initially, much of the ISOBus standard was based on protocols developed by the automotive industry; however, revisions have been made to support applications in the agricultural and forestry equipment industres. The main goal of ISO 11783 was to standardize electronic communications between tractor components, implement components and the tractor and implement.

FlexRay is a distributed network protocol that has been developed to improve on existing CAN technology. These protocols are typically developed by the automotive industry, but are soon integrated into agricultural vehicles as was seen with the CAN protocol under ISO standard 11783. One of the problems associated with existing CAN protocols is that in some cases, manufacturers are coming to a point where bus capacity will be exceeded. FlexRay offers the ability for data to be transferred at higher frequencies (10Mbps) compared to existing CAN protocols (250kbps) typically used today (National Instruments, 2010). Another important aspect of FlexRay is that it utilizes a time-triggered protocol that allows data to be transmitted and received at predetermined time frames which helps to eliminate errors that can occur when multiple messages are sent out on the bus. Additionally, the FlexRay protocol is capable of operating as a multi-drop bus, star network, or hybrid (using both multi-drop and star) network. This allows the protocol to be adapted easily into existing bus protocols while also providing increased reliability where desired with the star network. As automotive and agricultural vehicles develop in the future, FlexRay will certainly be the next network protocol used to ensure efficient and reliable data communication.

Data structures – While on-vehicle communication has relatively well defined data structures (ISO 11783), standards for transfer of data between the farm office and field machinery continue to evolve. The latter is being driven for the most part by software developers who recognize the need to reconcile data transfer from the farm office to field machinery and back again. Today, the need to reconcile data is being driven by map-based application. "Prescription maps" direct where and how inputs will be applied to crop production systems. Data regarding input metering and placement is further complicated by the nature of field equipment apply inputs. Crop production managers and suppliers have multifaceted data transfer needs that range from moving prescription maps form the farm office to field equipment and then returning plans field operations verification files along sensor data for summarizing crop health and performance to the field office.

One attempt at coordinating data transfer has been proposed and adopted by Macy (2003) and is termed the Field Operations Data Model (FODM). FODM was created as a framework to document field operations, and more recently has been expanded to support business functions. FODM is based on three components: description of field operation, framework and a general machine model. Field operations are described using one of four models; whole-field, product-centric, operations-centric of precision ag. The FODM framework is object-based which includes resources (people, machines, products, and domains) and operation regions (space and time). Data logged to summarize field operations can either be infrequently changing data (ICD) of frequently changing data (FCD). The general machine model (GMM) provides a description of the physical features of field machines including components, sensors and product storage or containers. An example of a machine definition using the GMM is shown in figure.

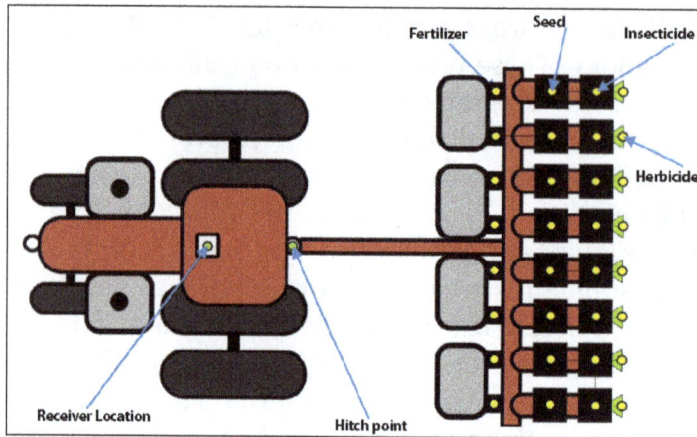

Illustration of a machine definition for a tractor/planter combination with multiple product metering and delivery systems using the GMM.

Automated guidance - Systems designed to accomplish automated guidance on agricultural vehicles can be seen back as far as the 1920s when furrows were used to guide tractors across fields with reduced effort from the operator. Since that time, as technology has improved, automated guidance has evolved from mechanical sensing to electronic sensors, machine vision, and GPS to successfully navigate equipment across the field. In most cases, operators utilize automatic guidance to follow parallel paths through the field. At the beginning of field operations, an A-B line is input into the control console, and the GPS coordinates are stored. As the operator continues to cover the field, the automatic guidance system can be engaged and the equipment will attempt to follow parallel paths to cover the field based on steering sensor feedback and GPS data. Many systems also provide the ability to follow curved paths which are input in much the same way.

Two basic types of automated guidance systems are typically used today by producers. The first system consists of a steering actuator which is mounted to the tractor's steering wheel. The second system is integrated into the tractor's steering system and utilizes a control valve to actuate the hydraulic steering cylinder directly. The overall accuracy of these systems relies heavily on the type of GPS technology used (RTK GPS provides the highest accuracy) as well as proper installation and setup. Ultimately, these systems benefit producers by reducing operator effort and pass-to-pass overlap during field applications.

Automated turns - After the successful development and employment of automated guidance on agricultural vehicles, the next logical step was to automate turning maneuvers. Creating a control system to automate turning at headland areas depends on several factors including headland width, equipment width, tractor dynamics, and the type of turn desired. One system that is currently available to producers which can automate turning movements is from Deere and Co. The iTEC Pro system uses tractor and equipment parameters and headland boundaries input by the operator to develop appropriate headland turns. Once engaged, the system will automatically perform the headland turn once it has entered a headland area without any input from the operator. An additional function provided by iTEC Pro is implement control. Control sequences can be setup for the equipment as it enters and exits the headland area. For instance, as the equipment enters the headland, tractor speed may be reduced, the implement raised. As the implement exits the headland, the implement may be lowered and the tractor speed increased. Using these two functions included in the iTEC Pro system; headland turns can be completely automated such that the operator does not need to steer the tractor nor activate the implement being utilized.

Harvest automation - Over the past two decades, yield monitors have been one of the most significant developments in harvesting technology. Manufacturers continue to improve these systems to provide yield and moisture measurements to the operator during harvesting operations as well as computer software for post-processing. Many producers utilize using automatic steering systems on harvesters to improve field efficiency. Most systems rely on GPS for guidance, however, systems have been developed which sense the stalks at the header to improve automated steering while harvesting corn. Improving grain quality and reducing grain loss is another method that producers can use to increase overall harvest efficiency. The development of hillside harvesters actually helped to improve cleaning capacity on steeper slopes, and harvesters are now offered by manufacturers including Deere and New Holland which have self-leveling cleaning shoes.

Future Trends in Automation

Liability – Perhaps one of the major impediments to development of fully autonomous field machinery is liability, and more important is who will assume or share the liability. For the foreseeable future, tractors will have drivers who in actuality are being relegated to baby sitters to a large extent because of equipment size and corresponding power levels. In short technology continues to remove much of the control responsibility from the operator. Perhaps the best examples include automated guidance and turns. Now on the horizon is automation of the combine threshing mechanism and cleaning shoe. Until manufacturers and producers reach a consensus as to how liability issues will be resolved, we can expect the operator to transition from commanding single machines to responsibility for multiple machines working in a coordinated behavior.

Use of multiple machines for increasing rate of work and productivity is common on most of the large scale farms worldwide. In a setup where multiple machines are used for agricultural production, one operator is required for each machine resulting in a one to one ratio of human operators to number of machines. Row crop operations like grain harvesting require at least two machines with one operator for each machine. The capability to manage and monitor both the harvester and the wagon by one operator can increase the field efficiency and reduce labor costs drastically. Algorithms for operating a master-slave multi-robot system were developed by. In this system the master machine is controlled manually and the autonomous slave machine has the capability to either follow or go to a particular location as commanded by the master machine. Vougioukas proposed a method for coordinating teams of robots where one master machine specifies the motion characteristics of one or more machines (slaves). Although no experiments were done with the proposed method, the simulation experiments verified that the method can be used for coordinated motion of hierarchies of master-slave robots.

The transition to fully autonomous operation will include a progression that begins with smaller, low power machines operated in controlled settings. When possible, fences or natural barriers might be utilized to corral errant vehicles. Lowenberg-DeBoer recognized this possibility when he concluded "Autonomous farm equipment may be in our future, but there are important reasons for thinking that it may not be just replacing the human driver with a computer. It may mean a rethinking of how crop production is done. In particular, once the driver is not needed, bigger is no longer better. Crop production may be done better and cheaper with a swarm of small machines than with a few large ones."

First generation unmanned machines - First generation unmanned machines are autonomous machines that require constant supervision despite the fact that they are autonomous. These machines lack the intelligence to cope with circumstances that are unexpected and dynamic. In the event of an emergency, the autonomous machine will either stop completely or alert a remote supervisor to aid it in mitigating the emergency.

Researchers at Carnegie Mellon University developed an autonomous harvesting machine known as Demeter system. The robotic machine harvested more than 40 hectares of crop without human intervention. The base machine was a retrofitted New Holland 2550 self- propelled windrower. Researchers at the Technical University of Denmark developed an autonomous robot prototype specifically for weed mapping. This robot was developed to mitigate the adverse effects of weed species like waterhemp that are developing glyphosate resistance. French and Spanish institutions in collaboration with equipment manufacturers developed a citrus harvesting robot (IVIA, 2004). This robot is different from weeding or scouting robots as it has an on-board manipulator to identify and harvest citrus fruit. Similar research efforts to develop citrus harvesting robots were conducted at the University of Florida by Hannan et al.

Robotic harvesters for specialty crops like cherry tomatoes, cucumbers mushrooms, cherries and others fruits have also been developed.Although, autonomous robotic manipulators are commercially available for milking and horticultural applications, mobile field robots are still not commercially available. The most sophisticated tractors available today feature automation of numerous machine functions but, require an operator to closely monitor the tasks being performed.

John Deere Company is currently working on a project to enable a single, remote user to supervise a fleet of semi-autonomous tractors mowing and spraying in an orchard. In a similar effort, three autonomous peat harvesting machines performed 100 field test missions during tests conducted with end users. The successful implementation of a multi-robot system by these researchers is a testimony to the fact that Ag-robots can work in real-world applications and the field of agriculture is evolving in to a high-tech work environment. Although autonomous, these first generation systems require close supervision by human operators and require further improvements to transform them into intelligent autonomous machines.

Individual robot control architectures – Most of the initial work done on control architectures of mobile robots was carried out in the aerospace and artificial intelligence research laboratories to accomplish military missions and space explorations. Unlike industrial robots, where the environment is controlled and structured, the work environment of Ag-Robots is relatively unstructured, unpredictable and dynamic. An intelligent, robust and fault tolerant control architecture is essential to ensure safe and desired operation of the Ag-Robot. A behavior based (BB) control approach provides an autonomous mobile robot, the intelligence to handle complex world problems using simple behaviors. Complex behaviors of a robot emerge from simple behaviors, behavior being defined as response to a stimulus. BB control structure can be either reactive or deliberative in nature. Reactive behaviors are part of reactive control architectures where the behavior responds to stimuli and develops control actions. Deliberative behaviors on the other hand are pre-defined control steps which are executed to accomplish a given task. Associating these behaviors to actual actions of an agricultural robot is crucial to understand the capabilities of a robot. The importance of decomposition of agricultural tasks into robotic behaviors was illustrated by Blackmore et al. For the robot to tackle unknown environments and attain assigned goals both reactive and deliberative behaviors are important and thus a robust fault tolerant intelligence is achievable with a combination of reactive and deliberative behaviors.

An Autonomous Robot Architecture (AuRA) for reactive control was developed by Arkin. Arkin mentioned three important aspects of a successful multi-purpose robot; motor behaviors that are used to describe the set of interactions the robot can have with the world, perceptual strategies that provide the required sensory information to the motor behaviors, and world knowledge both a *priori* and acquired that are used to select the motor behaviors and perceptual strategies that are needed to accomplish the robot's goals. AuRA consists of five basic subsystems; perception, cartographic, planning, motor and homeostatic control. Yavuz and Bradshaw did an extensive literature review of the available robot architectures and proposed a new conceptual approach to the design of hybrid control architecture for autonomous mobile robots. In addition to reactive, deliberative, distributed and centralized control approaches, fuzzy logic and modular hierarchical structure principles were utilized. Thus, three types of control architectures were acknowledged in the literature; hierarchical or deliberative, reactive and hybrid. The computability and organizing principles for each architecture differs and have their own peculiar set of building blocks. Essentially all BB architectures are software frameworks for controlling robots. BB robotic systems are significant, in the case where the real world cannot be accurately modeled or characterized. Uncertain, unpredictable and noisy situations are inherent characteristics of an agricultural environment and hence utilizing BB robotic architecture principles may be ideal.

A specification of behavioral requirements for autonomous tractor was provided by Blackmore et al. The authors discussed the importance of a control system that behaves sensibly in a semi-natural environment, and identified graceful degradation as a key element for a robust autonomous vehicle. Using the BB robotic principles, Blackmore et al developed a system architecture for the behavioral control of an autonomous tractor. Blackmore followed the assumption that robotic architecture designs refer to a software architecture, rather than hardware side of the system. In a more practical approach, a system architecture that connects high level and low level controllers of a robotic vehicle was proposed by Mott et al. In addition to the aforementioned levels, a middle level was introduced to improve the safety of the autonomous vehicle. The middle level enforced timely communication and provided consistent vehicular control. When the high level was not transmitting appropriately, the middle level recognizes this condition and transitions to a safe mode where the vehicle shuts down and stops. Ultimately, the middle level acts as a communication bridge integrating the high and low level controllers providing robustness to the robotic vehicles. This concept was successfully deployed on a fully-autonomous stadium mower and a large-scale peat moss harvesting operation.

Multi-robot control architectures – Coordinating multiple autonomous robots for achieving an assigned task presents an engineering challenge. When multiple robots are working together to accomplish a task the foremost question to be resolved is the type of inter-robot communication required. Inter-robot communication forms the backbone of a MRS. Identifying the specific advantages of deploying inter-robot communication is critical as the cost increases with the complexity of communication among the robots. Three types of inter-robot communication were explored by Balch et al. They found that communication can significantly improve performance in some cases but for others, inter-agent communication is unnecessary. In cases where communication helps, the lowest level of communication is almost as effective as the more complex type. Rude et al developed a wireless inter robot communication network called IRoN. The two important concepts of the network were implicit and explicit communications. A modest cooperation between robots is realized using implicit communication and a dynamic cooperation is achieved by using explicit communication. The authors utilized two robots to implement IRoN and were able to identify the changes which reduced the motion delay time ranges from 1000 ms to 50 ms. Wilke and Braunl developed flexible wireless communication network for mobile robot agents. The communication network was an explicit communication method which was applied to team members of a RoboCup team playing soccer. The communication network allowed broadcasting, transmission of messages between individuals and communication with a remote computer workstation. Fung et al. utilized a wireless transmitter and receiver to communicate position data. In their approach the position of a robot is gathered from infra-red sensor data and then transmitted to other robots via a radio link. The communication network is dedicated to sending only infra-red sensor data which makes it an inflexible network. A sophisticated technique called Carrier Sense Multiple Access/Collision Detection (CSMA/CD) was developed by Wang and Premvuti. The CSMA/CD protocol allows wireless inter-robot communication among multiple autonomous vehicles with a centralized supervisor.

To date, most of the research work done on multi-agent robot systems has been conducted in areas other than agriculture. Research work done on the architectural specifications of a MRS specifically deployed for agricultural production is rarely found in the literature. Thus, there is a need to understand, explore and research the control methodologies of a MRS so that multiple Ag-Robots can be deployed for agricultural production. Furthermore, the rapidly evolving contemporary agriculture industry may be poised to adopt MRS for increasing production efficiency

Next generation of autonomous field machinery – The next generation machines can be envisioned to accomplish agricultural production tasks autonomously using the intelligence provided by robust control architectures. As an example, two autonomous vehicles are assumed to perform baling and bale moving operations. Establishing communication between the baler and bale spear vehicles, hay bale location identification, navigation to the bale, spearing of the bale and relocation to the edge of the field will be done with minimal human supervision. Momentary wireless communication is established between the baling and bale spear vehicles during the spearing operation. The baling vehicle sends the location where it dropped the hay bale to aid the bale spear vehicle in path planning. The baling and spearing vehicles each have message frames to communicate the status and location of the bale. When bale is ejected, the vehicle transmits the location and timestamp through the Tx-message frame to the spear. The information about the bale is received by the Rx-message frame of the spearing vehicle which acknowledges the reception by transmitting a Tx-message frame. The baling and spearing vehicles, in addition to point to point communication, broadcast their messages with information containing their unique IDs, states, time stamp and the status of the assigned work to Central Monitoring Station (CMS).

In another instance, three Ag-robots are assumed to be a Combine, Grain Cart I and Grain Cart II.

Grain Carts I and II (followers) receive instructions from the Combine (leader) to navigate t along specific paths to off-load the harvested grain. Continuous point to multi-point communication between the Combine and Grain Carts is established. Grain Cart I and II maintains their trajectories at (b, o) and (b, L) relative to the trajectory of the leader for receiving the harvested grain. In addition to point to multi-point communication the states of the all the vehicles are broadcasted to the CMS.

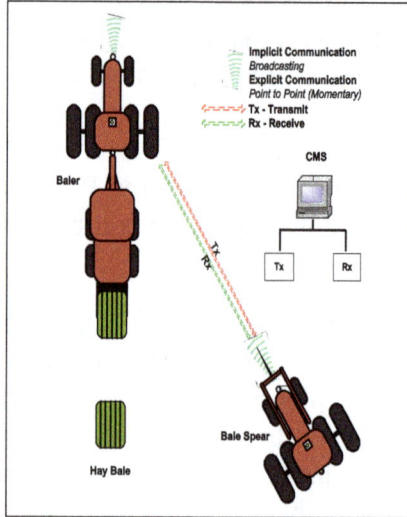

Coordinated vehicle navigation for performing point to point retrieval operations.

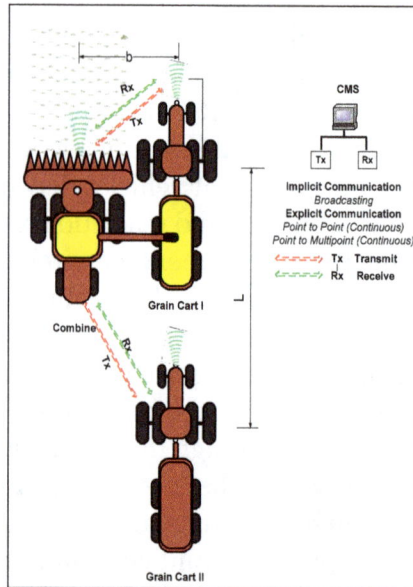

Coordinated vehicle operation for accomplishing biomass harvest, accumulation and transfer operations (leader–follower behavior).

In a more complex multi-robots system behavior instance, four autonomous vehicles simultaneously seed the same field. The Vehicles divide the seeding task into multiple working zones and perform work in their own zones. The control architecture provides intelligence to the seeding vehicles that divide the task and delegate specific vehicles to work in their own zones. The autonomous seeding vehicles broadcasts messages with information containing their unique IDs, states, time stamp and the status of the assigned work to the CMS. Each vehicle is assigned a unique ID. The status of work in this case would be the percentage of total area seeded by each vehicle. The CMS receives the data and stores all the data in its database for monitoring and post processing.

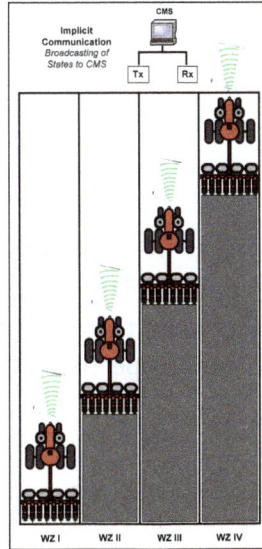

Coordinated navigation of multi-vehicle system for accomplishing a production task such as planting.

Equipment used in Harvest Automation

Grain Cart

A grain cart, also known as a chaser bin, is a trailer towed by a tractor. The term grain cart is used to represent a tractor-trailer system. Figure shows the appearance of a grain cart. Because of the larger capacity, one can use it to collect grains from multiple combines and transport them to a nearby truck or depot.

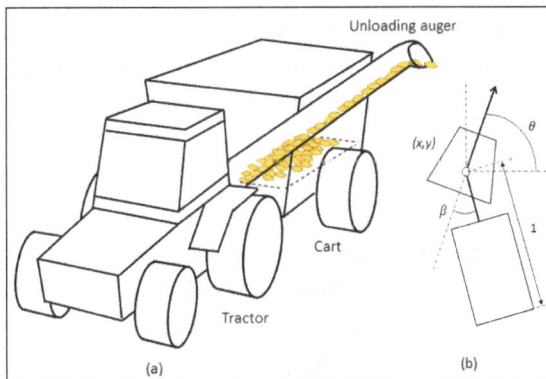

Grain cart.

Figure shows a grain cart. We model the grain cart as a trailer attached to a car-like robot. The robot is hitched by the trailer at the center. The length of the rotation arm for the trailer is assumed to be 1. The equations of motion for the grain cart are as follows:

$$\dot{q} = \begin{pmatrix} \dot{x} \\ y \\ \dot{\theta} \\ \dot{\beta} \end{pmatrix} = \begin{pmatrix} v\,cos\theta \\ v\,sin\theta \\ \omega \\ -v\,sin\beta + \omega \end{pmatrix}$$

Where $q = (x, y, \theta, \beta) \in \mathbb{R}^2 \times \mathbb{S}^1 \times \mathbb{S}^1$ is the configuration and $u = (v, \omega) \in U = [-1,1]^2$ is the control. In the configuration q, (x, y) is the coordinate of robot's center, θ is the robot's orientation, β is the angle between tractor and the trailer. v and ω denote the speed and angular velocity of the robot, respectively.

Problem Formulation

Harvesting Operation.

At the beginning of the harvesting operation, the combine that harvests the row nearest to the depot enters the field. Subsequently, the rest of the combines enter the field sequentially. While serving a combine, the grain cart is always on the side which is closer to the depot to minimize the distance traveled. To avoid collision with the combines, when a grain cart meets a combine on its way to the depot, it waits for the combine to move first. When a grain cart is filled, it travels to the depot to unload itself. As shown in Figure, we number the combines and the grain carts in the order in which they enter the field. We assume that all the combines have the same tank capacity C, filling rate r_f and unloading rate r_u. In the beginning, the combines enter the field along adjacent rows sequentially with constant velocity V_c and a time gap ΔT. Each grain cart serves a combine till it is completely empty. Then it moves to the next combine. We introduce the following notations to denote the pertinent time intervals during this operation:

- T: Time for each grain cart to move between two adjacent combines.

- $T_f \left(= \dfrac{C}{r_f} \right)$: Time for a combine to fill its tank.

- $T_u \left(= \dfrac{C}{r_u - r_f} \right)$: Time for a grain cart to empty the tank of a fully filled combine.

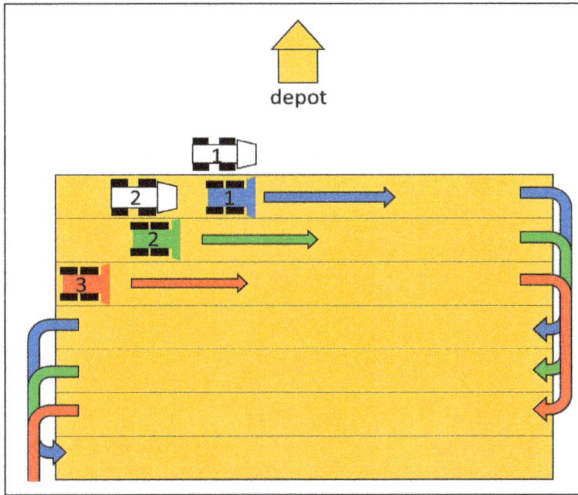

Motion of grain cart between combines.

When the harvesters reach the end of the field, they will stop moving and a grain cart will empty all the combines. Figure shows the sequence of operations for the combines. After harvesting a row, the combines move to the next row. Each row of harvesting starts with a new serving cycle. The turning trajectory at the end of each row is based on the edge condition which has been analyzed in Jin and Tang. The authors present trajectories for a combine to travel from one row to another with the smallest overlap or shortest path distance based on different edge conditions.

We make the following assumptions. (i) The farmland has a constant crop density. Initially, the analysis is for the simple case of a rectangular field which is subsequently generalized to the case of non-rectangular fields (ii) All the combine harvesters have the same tank capacity and forward velocity. (iii) The traveling time of the grain cart between combines i and j is $|i-j|T$.

Unloading Scheduling Strategy with Single Grain Cart

We first consider the problem of scheduling a single grain cart to serve N combines. Once the grain cart gets filled, it goes to the depot to unload the grains. We assume that the capacity of the grain cart is not sufficient to hold the grains in the entire field.

The combines enter the field sequentially with a constant time gap ΔT. We number the combines 1, 2,...,N in the order they enter the field. The grain cart serves the combines in the same order. After all the combines are served once, the grain cart goes to the depot and unloads the grains. The capacity of the grain cart is $C_g = NC \dfrac{r_u}{r_u - r_f}$. Let T_d denote the time required by the grain cart to make a trip to the depot before it arrives at the next combine. In general, T_d varies due to the change in the relative distance of the combine from the depot during the harvesting process.

Lemma. T_d is a constant for the strategy that minimizes the distance traveled by the grain cart during its trip to the depot. T_d is equal to the longest time taken by the grain cart to travel to the depot after serving a combine.

Proof. For a static depot, the distance between the grain cart and the depot changes during the harvesting operation. As the distance between the combines and the depot reduces during the harvesting operation, the grain cart can visit the next combine earlier. Let $C' \leq C$ be the volume of the load in the tank of the combine when the grain cart arrives. The time required for the grain cart to fill its tank to C_g is : T_2

$$= \frac{C'}{r_f} + \frac{C'}{r_u - r_f} + \frac{C_g - C'_g}{r_f} = \frac{C}{r_f} + \frac{C}{r_u - r_f}, \quad \text{where } C'_g = \frac{r_u}{r_u - r_f} C'. \text{ Therefore, the total time}$$

spent by a combine to unload an amount C_g is independent of C' :

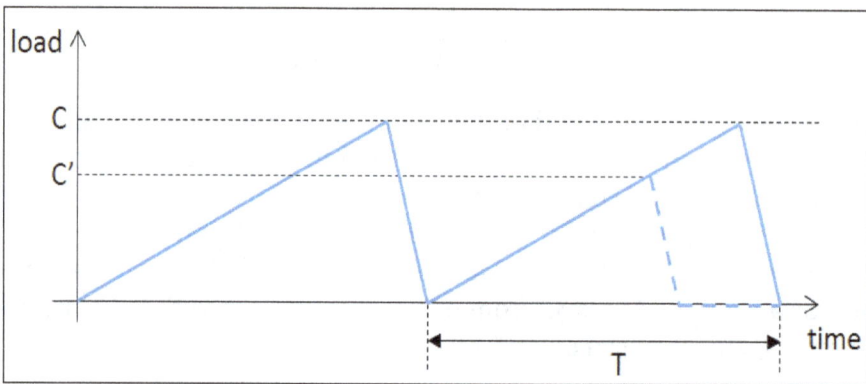

Load variation.

Figure shows the path followed by the grain cart between the time it leaves the depot and the time at which it unloads the combine to fill its own tank. If it arrives at the combine when its tank is filled at a level $C' < C_g$, it follows path L_2 to reach the combine and then travels along path L_3 to fill its tank by unloading the combine. The other option for the grain cart is to arrive at the combine just when its tank is completely full following path L_1. This may require it to wait at the depot after it finishes unloading itself. Since $L_1 < L_2 + L_3$, the second option is better for the grain cart to minimize the distance traveled. Therefore, the grain cart will wait at the depot in case it finishes unloading early in order to arrive at the combine just when its tank is completely filled.

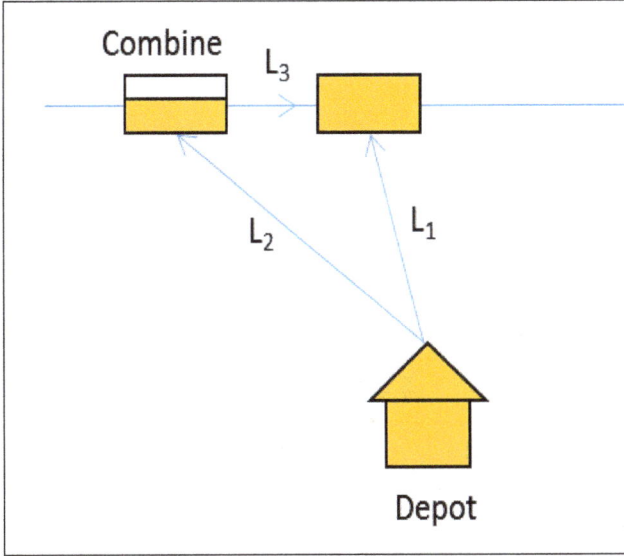

Traveling path of grain cart.

Since T_d is a constant, we assume $T_d = \lambda T$ (λ is a scaling parameter). The time during which a single combine fills its tank equals to the time for a grain cart to serve all the other combines and unload itself. Therefore, we can obtain the expression:

$$(N-1)\left(T + \frac{C}{r_u - r_f}\right) + \lambda T = \frac{C}{r_f}$$

$$\Rightarrow C = \frac{(N-1+\lambda)(r_u - r_f)r_f T}{r_u - Nr_f}$$

From Equation $(N-1)...\dfrac{(N-1+\lambda)(r_u - r_f)r_f T}{r_u - Nr_f}$, we can obtain the minimum required capacity C of the combines to carry out the harvesting operation with the given number of vehicles. This equation gives us the relationship between all the parameters involved in a harvesting operation when there are multiple combines and one grain cart. In addition to the minimum capacity of the combines, we can also obtain other parameters from $(N-1)...\dfrac{(N-1+\lambda)(r_u - r_f)r_f T}{r_u - Nr_f}$. For example, given C, r_f and r_u, we can calculate N.

Figure shows the load variation of the vehicles when $N = 2$ and $M = 1$.

Optimal Depot Location

In the scheduling strategy proposed one question may be asked is that where the depot should be located to minimize the travel distance of the grain cart.

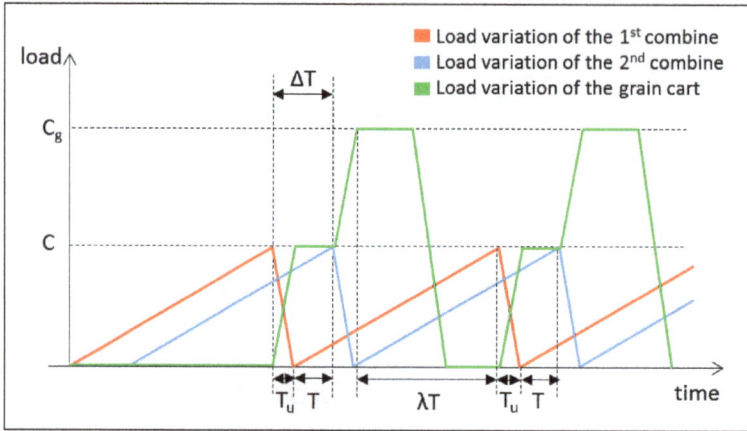

Load variations of 2 combines and 1 grain cart.

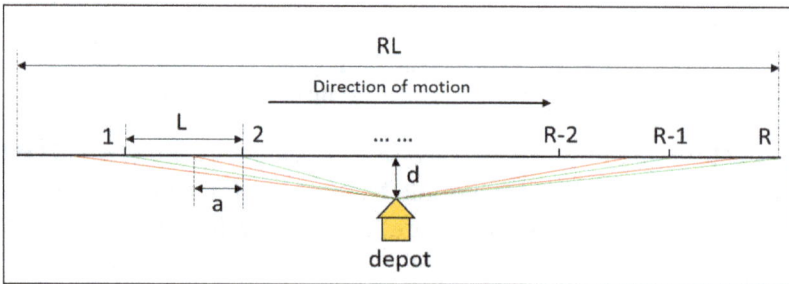

Working path of the vehicles.

We assume that the grain cart unloads N combines and visits the depot located next to the field. Then it returns to the first combine to repeat the process. We call this a cycle. During each cycle, the combines travel a constant distance L since the field is assumed to have a constant crop density. We assume R complete cycles are performed while harvesting a complete row which implies that the length of the row is RL. During the time the grain cart visits the depot, the combines travel for a distance a ($a < L$). As shown in Fig, the grain cart follows the red path to go to the depot and comes back along the green one in each cycle. We can see that the length of each path is not the same which leads to a changing traveling time. We assume that the grain cart follows an optimal strategy to minimize the distance traveled during its visit to the grain cart. From Lemma 1, this leads $T_d \left(T_d = \lambda T \right)$ to be a constant. The grain cart reaches the first combine exactly at the time it just gets filled.

To reduce the fuel and time spent during the harvesting operation, we need to find the optimal position of depot so that the total distance traveled by the grain cart is minimized. First, we consider the scenario of a single row. In this case, each combine harvests from one end of the field to the other only once to finish the harvesting operation.

Single Row

Before the combines reach the end of row, the grain cart goes to the depot N times which means the total length of the field is RL. Assume that the horizontal distance between the depot and the beginning of the row is xL, where $1 < x < R$, $x \in \mathbb{R}$ and the vertical distance between the depot and field is d, $0 \leq d$, $d \in \mathbb{R}$. As shown in Fig, before the combines reach the depot, for any period n, the distance traveled by the grain cart is $\sqrt{((x-n)L)^2 + d^2} + \sqrt{((x-n)L+a)^2 + d^2}$. After the combines pass the depot, it becomes $\sqrt{((n-x)L)^2 + d^2} + \sqrt{((n-x)L+a)^2 + d^2}$, where a is the distance traveled by the grain cart during the time when the grain cart is leaving. Therefore, the total distance traveled by the grain cart L_{total} can be expressed as follows:

$$L_{total} = \sum_{n=1}^{\lfloor x \rfloor} \sqrt{((x-n)L)^2 + d^2} + \sqrt{((x-n)L+a)^2 + d^2} +$$

$$\sum_{n=\lfloor x \rfloor+1}^{R} \sqrt{((n-x)L)^2 + d^2} + \sqrt{((n-x)L-a)^2 + d^2}$$

The problem is to find the optimal position of the depot so that the total distance traveled by the grain cart is minimized.

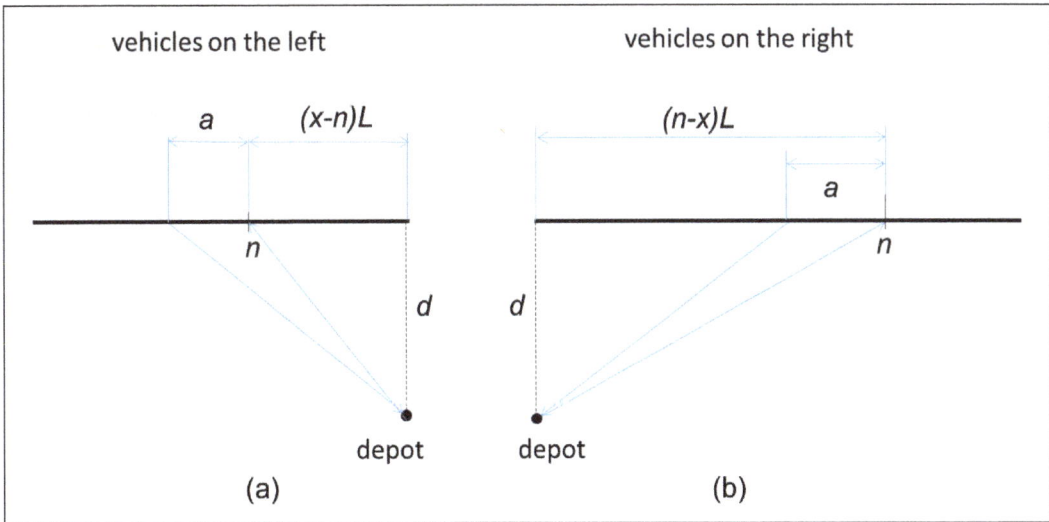

Travel Condition.

Proposition: The optimal position of the depot is $x^* = \dfrac{(R+1)L - a2}{2}$.

Proof. As shown in Figure, the grain cart visits the depot N times while harvesting the first row. We define the path followed by the grain cart to visit the depot for the nth time and come back, as the nth path. We divide the n paths into pairs. We let the nth path and the $(R+1-n)^{th}$ path be a pair. If R is even, there are $\dfrac{R}{2}$ pairs. When R is odd, there are

$\frac{R-1}{2}$ pairs and the one in the middle, which is $\left(\frac{R+1}{2}\right)^{th}$ path is unpaired. The length of each pair of paths is given by the following expression:

$$L_{pair} = \sqrt{\left((x-n)L\right)^2 + d^2} + \sqrt{\left((x-n)L+a\right)^2 d^2}$$

$$+ \sqrt{\left((R+1-n-x)L\right)^2 d^2} + \sqrt{\left((R+1-n-x)L-a\right)^2 + d^2}$$

The second derivative L''_{pair} is:

$$L''_{pair} = \frac{L^2 d^2}{\left(\left((x-n)L+a\right)^2 + d^2\right)^{\frac{3}{2}}} + \frac{L^2 d^2}{\left(\left((x-n)L\right)^2 + d^2\right)^{\frac{3}{2}}}$$

$$+ \frac{L^2 d^2}{\left(\left((x-1)L\right)^2 + d^2\right)^{\frac{3}{2}}} + \frac{L^2 d^2}{\left(\left((x-1)L+a\right)^2 + d^2\right)^{\frac{3}{2}}} \geq 0$$

Since $L''_{pair} > 0$, L_{pair} is a convex function Boyd and Vandenberghe. Therefore, the minimum exists and the solution is unique. L'_{pair} implies $x^* = \frac{(R+1)L-a}{2}$. This is true for all the pairs of trajectories, so we know L_{total} is also convex and reaches its minimum at $x^* \frac{(R+1)L-a}{2}$. When R is odd, the $\frac{R+1}{2}$th path has no other group to pair with. Its length is $L_{mid} = \sqrt{\left(\left(x-\frac{R+1}{2}\right)L\right)^2 + d^2} + \sqrt{\left(\left(x-\frac{R+1}{2}\right)L-a\right)^2 + d^2}$. It can be shown that L_{mid} is also convex and reaches minimum at $x^* = \frac{(R+1)L-a}{2}$. The proposition is proved.

Multiple Rows

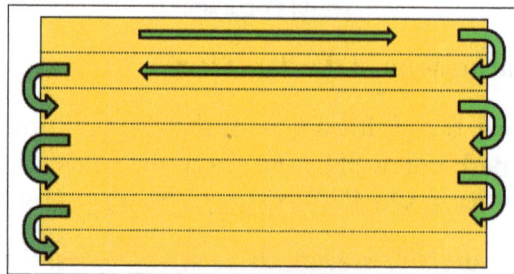

Order of harvesting operation.

When the field contains r rows, the combines harvest from one end of the field to the other. After finishing the current row, the combines turn to the nearest unharvested row. Fig. shows the order of harvesting operation. We notice that the direction of motion of the combines in the current row is always opposite to the previous one. We assume all the rows have the same width w. The distance traveled by the grain cart in any row i can be expressed as follows:

$$when\, i\, is\, odd: L_i = \sum_{n=1}^{\lfloor x \rfloor} \sqrt{\left((x-n)L\right)^2 + \left(d+(i-1)w\right)^2} \cdot$$

$$+\sqrt{\left((x-n)L+a\right)^2 + \left(d+(i-1)w\right)^2} +$$

$$\sum_{n=\lfloor x \rfloor}^{R} \sqrt{\left((n-x)L\right)^2 + \left(d+(i-1)w\right)^2} + \sqrt{\left((n-x)L-a\right)^2 + \left(d+(i-1)w\right)^2}$$

$$when\, i\, is\, even: L_i = \sum_{n=1}^{\lfloor R-x \rfloor} \sqrt{\left((R-x-n)L\right)^2 + \left(d+(i-1)w\right)^2}$$

$$+\sqrt{\left((R-x-n)L+a\right)^2 + \left(d+(i-1)w\right)^2} +$$

$$\sum_{n=\lfloor R-x \rfloor}^{R} \sqrt{\left((n-(R-x))L\right)^2 + \left(d+(i-1)w\right)^2} + \sqrt{\left((n-(R-x))L-a\right)^2 + \left(d+(i-1)w\right)^2}$$

$$L_{total}^r = \sum_{i=1}^{r} L_i$$

L_{total}^r is the total distance traveled by the grain cart when there are r rows. We know that L 0̈0 total $\geq$ 0, so the minimum of L_{total}^r exists and is unique. By solving $\left(L_{total}^r\right)' = 0$, we can find the optimal location x^* which minimizes the total distance traveled by the grain cart. The expression of L_{total}^r is as follows:

$$when\, i\, is\, odd: L_i' = \sum_{n=1}^{\lfloor x \rfloor} \left(-\frac{L^2(x-n)}{\sqrt{\left((x-n)L\right)^2 + \left(d+(i-1)w\right)^2}}\right.$$

$$+\frac{L\left(L(x-n)+a\right)}{\sqrt{\left((x-n)L+a\right)^2 + \left(d+(i-1)w\right)^2}}\right)$$

$$+ \sum_{n=\lfloor x \rfloor}^{R} \left(-\frac{L^2(n-x)}{\sqrt{((n-x)L)^2+(d+(i-1)w)^2}} - \frac{L(L(n-x)-a)}{\sqrt{((n-x)L-a)^2+(d+(i-1)w)^2}} \right)$$

$$when\, i\, is\, even: L_i' = \sum_{n=1}^{\lfloor R-x \rfloor} -\frac{L^2(R-x-n)}{\sqrt{((R-x-n)L)^2+(d+(i-1)w)^2}}$$

$$+\frac{L(L(R-x-n)+a)}{\sqrt{((R-x-n)L+a)^2+(d+(i-1)w)^2}}$$

$$+ \sum_{n=\lfloor R-x \rfloor}^{R} \frac{L^2(n+x-R)}{\sqrt{((n-(R-x))L)^2+(d+(i-1)w)^2}} + \frac{L(L(n+x-R)-a)}{\sqrt{((n-(R-x))L-a)^2+(d+(i-1)w)^2}}$$

$$L_{total}' = \sum_{i=1}^{r} L_i'$$

To obtain an estimate of x^*, we first obtain x_i^* for each row individually using the method presented. Then we can obtain the estimated optimal depot position $\overline{x^*}$ by calculating the mean value of the optimal location in each row, $\overline{x^*} = \frac{\sum_{i=1}^{r} x_i^*}{r}$. When i is odd, $x_i^* = \frac{(R+1)L-a}{2}$. When i is even, $x_i^* = \frac{(R-1)L+a}{2}$. Therefore, we can approximate the optimal depot position for r rows. When r is even, $\overline{x^*} = \frac{L-a}{2r} + \frac{RL}{2}$. When r is even, $\overline{x^*} = \frac{RL}{2}$. Next, we present a bound on the maximum error incurred in the approximation.

First, we consider the case when there are two rows. As before, we divide the paths into groups. We let the nth path and the $(R+1-n)^{th}$ path, $(n \leq \frac{R}{2})$ of each row be in group n. Each group contains 4, or 2 pairs of paths, one pair from each row. The length of paths in group n is given by the following expression:

$$L_{group} = \sqrt{((x-n)L)^2+d^2} + \sqrt{((x-n)L+a)^2 d^2}$$

$$+\sqrt{((n-x)L)^2+d^2} + \sqrt{((n-x)L-a)^2+d^2}$$

$$+\sqrt{\left((R-x-n)L\right)^2+(d+w)^2}+\sqrt{\left((R-x-n)L+a\right)^2+(d+w)^2}$$

$$+\sqrt{\left((n-(R-x))L\right)^2+(d+w)^2}+\sqrt{\left((n-(R-x))L-a\right)^2+(d+w)^2}$$

$$L'_{group}=\frac{L^2(x-n)}{\sqrt{\left((x-n)L\right)^2+d^2}}+\frac{L(L(x-n)+a)}{\sqrt{\left((x-n)L+a\right)^2+d^2}}-\frac{L^2(n-x)}{\sqrt{\left((n-x)L\right)^2+d^2}}$$

$$-\frac{L(L(n-x)-a)}{\sqrt{\left((n-x)L-a\right)^2+d^2}}$$

$$-\frac{L^2(R-x-n)}{\sqrt{\left((R-x-n)L\right)^2+(d+w)^2}}-\frac{L(L(R-x-n)+a)}{\sqrt{\left((R-x-n)L+a\right)^2+(d+w)^2}}$$

$$+\frac{L^2(n+x-R)}{\sqrt{\left((n-(R-x))L\right)^2+(d+w)^2}}+\frac{L(L(R-x-n)+a)}{\sqrt{\left((R-x-n)L+a\right)^2+(d+w)^2}}$$

It was proved that L_{pair} is convex. Since L''_{group} is the summation of L''_{pair} of each row, $L''_{group}\geq0$. Therefore, for 2 rows, the total travel distance L_{total}, which is the summation of the length of paths in all the groups, is convex as well. By substituting $x=\dfrac{RL}{2}$ we can get that $L'_{group}>0.$, which indicates the optimal position of the depot is to the left of $x=\dfrac{RL}{2}$ when there are two rows. When there are more rows, as long as r is even, it can be generalized that L'_{total} increases. So $\dfrac{RL}{2}$. For any row i individually, the optimal depot position x_i^* is located at $\dfrac{(R+1)L-a}{2}$ or $\dfrac{(R-1)L+a}{2}$. When there are multiple rows, it is obvious that $\dfrac{(R-1)L+a}{2}<x^*<\dfrac{(R+1)L-a}{2}$. So when there are even rows, we know that $\dfrac{(R-1)L+a}{2}<x^*<\dfrac{RL}{2}$.

When r is odd, the only difference is that there is always an additional row starting from $r=3$. By substituting $x=\dfrac{L-a}{2r}+\dfrac{RL}{2}$ to equation $L'_{total}=\sum_{i=1}^{r}L'_i$ we obtain $L'_{group}>0$. .r increases, L'_{group} increases. As a result, x^* is to the left of $\dfrac{L-a}{2r}+\dfrac{RL}{2}$ when r is odd.

In this case, $\dfrac{(R-1)L+a}{2}<x^*<\dfrac{L-a}{2r}+\dfrac{RL}{2}$. Therefore, we reach a conclusion that when r is even, the optimal location of depot is in $\dfrac{(R-1)L+a}{2}+\dfrac{RL}{2}$ while it is in $(\dfrac{(R-1)L+a}{2},\dfrac{L-a}{2r}+\dfrac{RL}{2})$ when r is odd. The maximum error is $\dfrac{(L-a)(r+1)}{2r}$.

Path Planning of the Grain Cart

Numerical Approach

The numerical approach is used to obtain the time-optimal lane change maneuver a description of the numerical approach to obtain the time-optimal solution for the grain cart to move between combines. We use q_i and q_g to denote the initial and goal configuration of the path. Here q_i is set to be the state when grain cart leaves the combine, and q_g is the configuration of the next combine.

Denote the set of admissible path from the configuration q_i as $\mathscr{A}\,x_i, y_i\,, \theta_i\,, \beta_i$. Given a goal configuration q_g, we define the corresponding value function $u : q \to \mathbb{R}^+ \cup \{0\}$.

$$u\big(q(T)\big) = \inf\big\{T : q(t) \in \mathscr{A}\,x_i\,, y_i\,, \theta_i\,, \beta_i\,, q(T) = q_g\big\}$$

The value function can be regarded as the optimal cost-to-go for the tractor-trailer model with given constraints, an initial configuration and a final configuration. By applying dynamic programming principle for the Eqn. ($u\big(q(T)\big) = \inf\big\{T : q(t) \in \mathscr{A}\,x_i\,, y_i\,, \theta_i\,, \beta_i\,, q(T) = q_g\big\}$), we have:

$$u\big(q(t)\big) = \inf\big\{u\big(q(t+\Delta t)\big) + \Delta t : q(t) \in \mathscr{A}\,x_i, y_i, \theta_i, \beta_i\big\}$$

Dividing the terms by Δt and taking $\Delta t \to 0$, we are able to derive:

$$-1 = \inf\big\{\nabla u\cdot\dot{q} : |v| = 1, |\omega| \le 1\big\}$$

With the equations of motion of the grain cart, Hamilton-Jacobi-Bellman equation is obtained as follows:

$$-1 = \cos\theta\,\frac{\partial u}{\partial x} + \sin\theta\frac{\partial u}{\partial y} - \sin\beta\frac{\partial u}{\partial\beta} + \inf_{|\omega|\le1}\{\dot{\theta}(\frac{\partial u}{\partial\theta} + \frac{\partial u}{\partial\beta})\}$$

The last term in Eqn. ($-1 = \cos\theta\,\dfrac{\partial u}{\partial x} + \sin\theta\dfrac{\partial u}{\partial y} - \sin\beta\dfrac{\partial u}{\partial\beta} + \inf_{|\omega|\le1}\{\dot{\theta}(\dfrac{\partial u}{\partial\theta} + \dfrac{\partial u}{\partial\beta})\}$) can be eliminated by applying bang-bang principle $w=\pm1$. Since q_i is the goal configuration which has no cost-to-go, we have $u(q_g) = 0$. For the points located in the obstacle or outside the space, we define the cost-to-go to be infinity.

Update Scheme

In order to find the time-optimal path satisfying Eqn. ($-1 = \cos\theta \frac{\partial u}{\partial x} + \sin\theta \frac{\partial u}{\partial y} - \sin\beta \frac{\partial u}{\partial \beta} + \inf_{|\omega| \le 1} \{\dot\theta(\frac{\partial u}{\partial \theta} + \frac{\partial u}{\partial \beta})\}$), we apply fast sweeping meth-od and propose an update scheme for the value function $u(q)$ for the entire space. The idea is to take advantage of the fact that the value function has zero cost-to-go at the goal configuration, and to compute the value function from the nodes close to the goal configuration, to the nodes at farther positions.

With this in mind, we first set up a four dimensional uniform Cartesian grid with re-finement $(h_x, h_y, h_\theta, h_\beta)$. Let $u_{a,b,c,d} = u(q_{a,b,c,d}) = u(ah_x, bh_y, ch_\theta, dh_\beta)$ be the approxi-mation of the solution on the grid nodes. Moreover, we discretize ω in the range of $[-1,1]$ and further define $u^*_{a,b,c,d}$ as follows:

$$u^*_{a,b,c,d} = \min_{\omega i \in [-1,1]} \{u(q_{a,b,c,d} + \dot{q}\Delta t)\} + \Delta t$$

where $\dot{q} = (\cos(ch_\theta), \sin(ch_\theta), \omega_i, -\sin(dh_\beta) + \omega_i)^T$, ω_i is the ith element in the dis-cretization and Δt is the length of time step. The value $u(q_{a,b,c,d} + \dot{q}\Delta t)$ is approximated by taking the average value of the adjacent nodes in the presented grid.

Finally, the update scheme can be described as follows:

$$u^{n+1}_{a,b,c,d} = \min\{u^n_{a,b,c,d}, \quad u^{*n}_{a,b,c,d},\}$$

, where the superscripts denote the iteration. We set up the termination condition of the computation as follows:

$$\left(\left\|u^{n+1}_{a,b,c,d} - u^{n+1}_{a,b,c,d}\right\|_2\right)^2 < \varepsilon$$

, where $\varepsilon > 0$.

Computing Trajectory

By using the obtained value function $u(q)$, we are able to derive the time-optimal path from any initial configuration q_i to the goal configuration q_g. The control law can be summarized as follows:

$$\dot{x} = \cos\theta$$

$$\dot{y} = \sin\theta$$

$$\dot{\theta} = -sgn\left(\frac{\partial u}{\partial\theta} + \frac{\partial u}{\partial\beta}\right)$$

$$\dot{\beta} = -\sin\beta + \dot{\theta}.$$

Note that the partial derivative in Eqn. ($= -sgn\left(\frac{\partial u}{\partial\theta} + \frac{\partial u}{\partial\beta}\right)$) is obtained by applying centered difference approximation. The values of u which are not on the nodes are computed using a nearest-neighbour interpolation.

The numerical approach computes the time-optimal trajectory efficiently if the corresponding value function is provided. The main time consumption is in computing the value function of the final configuration. But in real implementation, one can compute the value function beforehand. Hence, the time cost of computing the value function will not influence the real operation on path planning. Figure shows the time optimal path obtained with numerical approach of a grain cart moving between two points.

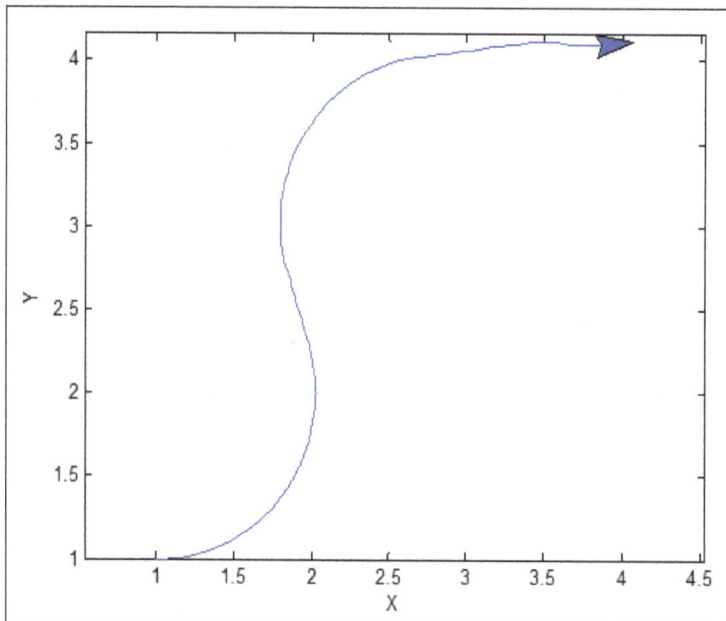

Path of grain cart with numerical approach.

Primitive-based Motion Maneuvre

The trajectory is generated by a numerical approach Zhang and Bhattacharya (2015) that is solution to find the time-optimal path between an initial and a final configurations for a tractor-trailer.

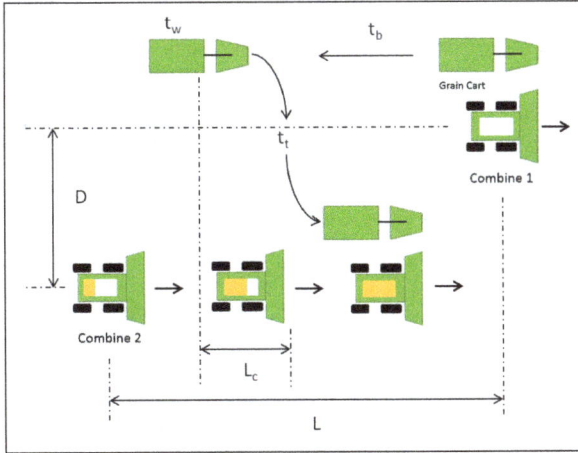

Path planning for the grain cart between two combines.

The proposed maneuver resembles a parallel parking maneuver for a tractor-trailer with the exception that the objective here is to "park" a grain-cart alongside a moving combine. Figure illustrates the position of the grain cart relative to the two combines as it implements the proposed maneuvers. Let L denote the horizontal distance maintained between two adjacent combines while they are moving on two adjacent lanes. The grain cart in Figure moves from combine 1 to combine 2. The trajectory contains the following primitives:

- Back-up: After serving the current combine, the grain cart moves backward in a straight line with a speed v_g while the combines keep moving forward with a speed v_c. Let t_b denote the duration of this phase.

- Stall: The grain cart stops, and waits for a time t_w until the relative position of combine 2 with respect to the grain cart is L_c as shown in the figure.

- Lane-change: The grain cart follows a time-optimal trajectory presented in Zhang and Bhattacharya (2015) to reach combine 2. This stage resembles a lane-change maneuver by vehicles observed frequently on highways. Subsequently, the grain cart moves parallel to the combine.

The only exception occurs when the grain cart returns to the initial combine after serving the combine at the other end. In this case, it simply performs the lane-change maneuver.

The maneuvre contains some decision variables, namely t_b and t_w. The aforementioned variables can be tuned to either minimize the distance traveled (D_{tb}) by the grain cart during the maneuvre or the total time spent $(T(t_b, t_w))$ to complete the maneuvre.

Based on the user's preferences, the following cost function can be formulated:

$$J(t_b, t_w) = \lambda T(t_b, t_w) + (1 - \lambda) D(t_b) \quad \lambda \in [0, 1],$$

where the parameter λ is decided by the user.

For a given distance D and L_c, let t_{lc} denote the time required to complete the lane-change maneuvre. This is computed off-line by the numerical approach. Based on the geometric constraints imposed by the maneuvre, we obtain the following equation:

$$L + L_c = \left(v_g + v_c\right)t_b + v_c t_w$$

$$\Rightarrow tw = \frac{\left(L + L_c\right) - \left(v_g + v_c\right)t_b}{v_c}$$

The expression of the total traveling time and distance is as follows:

$$T\left(t_b, t_w\right) = t_b + t_w + t_{lc}$$

$$\Rightarrow T\left(t_b\right) = t_b + \frac{\left(L + L_c\right) - \left(v_g + v_c\right)t_b}{v_c} + t_{lc}$$

$$D\left(t_b\right) = t_b v_g + t_{lc} v_{lc}.$$

Where v_{lc} is the velocity of grain cart during lane-changing. Substituting

$$J\left(t_b, t_w\right) = \lambda T\left(t_b, t_w\right) + (1-\lambda)D\left(t_b\right) \quad \lambda \in [0,1], \text{ in } J\left(t_b\right) = \left(\lambda - \frac{v_g + v_c}{v_c}\lambda + (1-\lambda)v_g\right)t_b.$$

$$+ \lambda t_{lc} + \left[\frac{L + L_c}{v_c} + (1-\lambda)t_{lc}v_{lc}\right] \text{ leads to the following expression for the cost function:}$$

$$J\left(t_b\right) = \left(\lambda - \frac{v_g + v_c}{v_c}\lambda + (1-\lambda)v_g\right)t_b + \lambda t_{lc} + \left[\frac{L + L_c}{v_c} + (1-\lambda)t_{lc}v_{lc}\right]$$

Since Jt_b is a linear function of t_b, we can conclude the following:

- At $\lambda = \dfrac{v_c}{v_c + 1}$ $j\left(t_b\right)$ is independent of t_b.

- For $\lambda < \dfrac{v_c}{v_c + 1}$, the minimum value of J_{tb} occurs at t_b = 0. This implies that the grain cart should wait after serving combine 1, and subsequently perform the lane change maneuvre to serve combine 2.

- For $\lambda > \dfrac{v_c}{v_c + 1}$, the minimum value of J_{tb} occurs at $t_b = \dfrac{L + L_c}{v_g + v_c}$. This implies that the grain cart should back-up after serving combine 1 till it is at a horizontal distance of L_c behind combine 2, and subsequently perform the lane change maneuvre. There is no time spent in the stall maneuvre since t_w = 0.

Therefore, the optimal value of tb lies at either end of the interval for $\lambda \neq \dfrac{v_c}{1+v_c}$.

Conveyor Belt

In today's world where transportation is perhaps one of the most elements for the business to succeed, people are always in search of better, faster and cost effective technique of transporting, tracking and processing their goods.

In the recent years, conveyor belt system has emerged as one of the best ways of transporting product. This system consists of two or more number of drums that are also referred as pulleys with a conveyor belt i.e. the carrying medium and it rotates around the drums. The belt is propelled by one or two of the pulleys. The pulley that powers the belt is called the drive pulley and the other is called idler pulley. The conveyor belt is categorized into two classes, one that transports general material and the other that is responsible for taking agricultural material such as sand, ore, coal, salt, grain and much more from one place to another.

Uses of Conveyor in the Agricultural Community

The uses of a conveyor belt in the agricultural community are numerous. It is used in the feeding process that processes or freezes vegetables, fruits, and other agricultural products for packing to be sent to the retailers or other process plants. Besides this, these belts are effectively being used for loading bulk quantities into trucks in order to get them transported to their final destination. Bean market is another place, where conveyor belt scales are used in a large number. Additionally, these belts are to load out the soy beans into the trucks that carry them to the end process.

These belts are designed to suit the requirements of the agricultural industry application. Sugar Cane, Cauliflower, peas, flour, rice, and grains are some of the agricultural applications that use conveyor belt scales. Other applications include:

- Seed coating for vegetable seeds and cereal grains,
- Chopping of crops,
- Dust collection,
- Harvesting,
- Threshing,
- Material handling,
- Storage,
- Grain processing,
- Drying of grains.

Merits of using the Belts in the Agricultural Industry

With the help of these conveyor belts, the workforce reduces and this leads to more profits and less expense. Also, the storage, drying, and processing of the products have become very efficient and easy. These processes are also easy to synchronize due to these belts. They have excellent turn-around time and handle the systematic materials very well.

These belts are specially designed in order to cater the needs of the agricultural world. They are not only cost effective, but also efficient. They have the ability to adjust to any type of temperature. Different conveyor belts are used to perform agriculture applications.

Grain Drying

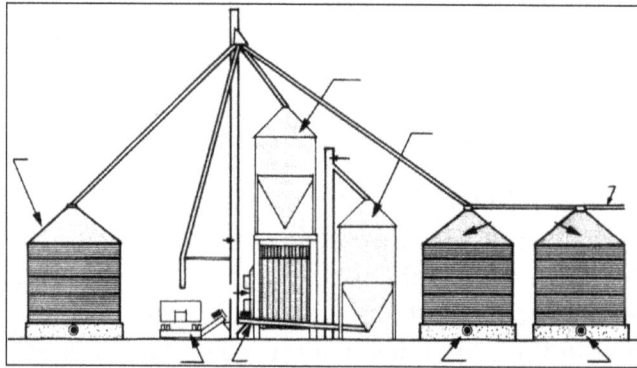

Grain drying, refers to the removal of some of the moisture from grain by mechanically moving air through the grain after it has been harvested. Grain in the field dries naturally as the crop matures, giving up moisture to the air until the grain moisture is in equilibrium with the moisture in the air (equilibrium moisture content). Conditions become less favorable for grain to dry to moisture contents considered safe for storage as the harvest is delayed into late fall.

Drying Advantages and Disadvantages

Grain drying has several advantages and disadvantages.

Advantages

- Increases quality of harvested grain by reducing crop exposure to weather.
- Reduces harvesting losses, including head shattering and cracked kernels.
- Reduces dependency on weather conditions for harvest.
- Allows use of straight combining for small grains.

- Reduces size and/or number of combines and other harvest-related equipment and labor required due to extending harvest time.

- Allows more time for post-harvest field work.

Disadvantages

- Original investment for drying equipment and annual cost of ownership.

- Operating costs for fuel, electricity and labor.

- Extra grain handling required may result in further investment for equipment.

Recommended Storage Moisture Contents and Estimated Allowable Storage Times

The length of time grain can be stored without significant deterioration is determined by temperature and the moisture content at which it is stored. Table 1 shows the maximum recommended moisture content for storage with aeration of some typical North Dakota grains. Short-term storage generally refers to storage under winter conditions while long-term storage considers the effect of summer conditions. Grain with damaged kernels or with significant amounts of foreign material needs to be stored at a 1 to 2 percentage points lower moisture content than sound, clean grain.

Grain can be stored at a higher moisture content without significant fungus development when stored at colder temperatures. Table shows the relationship between moisture and temperature and its effect on allowable storage time for cereal grains.

The allowable storage time for corn has been established to be the time until a 0.5 percent dry matter reduction is reached. At that point there will be a reduction of one grade. Storage life is cumulative. If half of the storage life is used before the grain is dried, only half of the indicated storage time at the lower moisture content is available after the grain has been dried.

Table: Maximum recommended moisture contents of selected clean, sound grains for storage with aeration in north dakota.

	Short term Long term (less than (more than 6 months) 6 months)	
Barley	14 %	12 %
Corn	15.5	13
Edible Beans	16	13
Flax seed	9	7
Millet	10	9

Oats	14	12
Rye	13	12
Sorghum	13.5	13
Soybeans	13	11
Non-Oil Sunflower	11	10
Oil Sunflower	10	8
Wheat	14	13

Table: "Approximate" Allowable Storage Time (days) For Cereal Grains.

M.C.	Temperature(°F)					
(%)	30°	40°	50°	60°	70°	80°
14	•	•	•	•	200	140
15	•	•	•	240	125	70
16	•	•	230	120	70	40
17	•	280	130	75	45	20
18	•	200	90	50	30	15
19	•	140	70	35	20	10
20	•	90	50	25	14	7
22	190	60	30	15	8	3
24	130	40	15	10	6	2
26	90	35	12	8	5	2
28	70	30	10	7	4	2
30	60	25	5	5	3	1

Based on composite of 0.5 percent maximum dry matter loss calculated on the basis of USDA research at lowa State University.

*Approximate allowable storage time exceeds 300 days.

A rough estimate of storage life for oil crops might also be made based on the values for corn using an adjusted moisture content calculated using the equation:

$$\text{Comparable Corn Moiisture Content} = \frac{\text{Oil Seed Moisture Content} \times 100}{100 - \text{Seed Oil Content}}$$

For example,oil sunflower at 12.0 percent moisture content is comparable to corn at 20 percent moisture content.

$$\text{Comparable Corn Moiisture Content} = \frac{12}{100 - 40} \times 100 = 20\%$$

Influence of Drying Conditions

Airflow rate, air temperature and air relative humidity influence drying speed. In general, higher airflow rates, higher air temperatures and lower relative humidities increase drying speed.

Raising the temperature of the drying air increases the moisture-carrying capacity of the air and decreases the relative humidity. As a general rule of thumb, increasing the air temperature by 20 degrees Fahrenheit (F) doubles the moisture-holding capacity of air and cuts the relative humidity in half.

The drying rate depends on the difference in moisture content between the drying air and the grain kernel. The rate of moisture movement from high moisture grain to low relative humidity air is rapid. However, the moisture movement from wet grain to moist air may be very small or nonexistent. At high relative humidities, dry grain may pick up moisture from the air.

The airflow rate also affects drying rate. Air carries moisture away from the grain, and higher airflow rates give higher drying rates. Airflow is determined by fan design and speed, fan motor size and the resistance of the grain to airflow. Deeper grain depths and higher airflow rates cause higher static pressures against the fan. Higher static pressures decrease fan output.

As air enters the grain, it picks up some moisture, which cools the air slightly. As air moves through a deep grain mass, the air temperature is gradually lowered and relative humidity increased until the air approaches equilibrium with the grain. If the air reaches equilibrium with the grain, it passes through the remaining grain without any additional drying. If high relative humidity air enters dry grain, some moisture is removed from the air and enters the grain. This slightly dried air will begin to pick up moisture when it reaches wetter grain. Air in a 12 to 16 inch grain column does not reach equilibrium with the grain.

Types of Dryers

Dryers can be categorized in different ways. There are natural air, low temperature, and high temperature dryers; there are batch, automatic batch and continuous flow dryers; and there are in-bin and column or self-contained dryers. Dryers can also be classified according to the direction of airflow through the grain; cross-flow, counter-flow, and concurrent-flow.

Natural Air/Low Temperature Drying

Advantages:

- No harvest bottle neck. The bins can be filled at the harvest rate.

- A properly sized system may dry the crop more economically than a high temperature dryer.

Disadvantages:

- There is a limit on initial moisture content that can be effectively dried.

- Electrical power must be available at each bin for dryer fan motors.

Natural air/low temperature drying refers to drying grain using little or no additional heat. Drying takes place in a drying zone which advances upward through the grain.

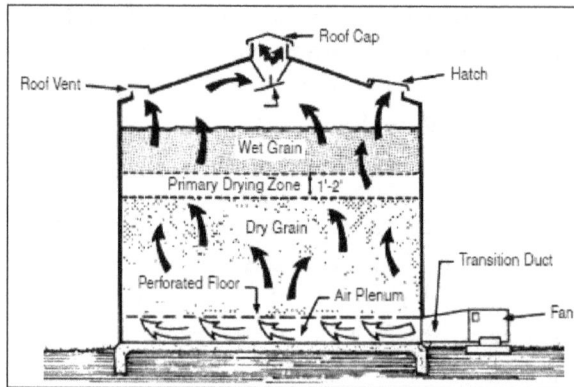

A typical bin dryer utilizing natural air/flow temperature drying.

Grain above this drying zone remains at the initial moisture content or slightly above, while grain below the drying zone is at a moisture content in equilibrium with the drying air. The equilibrium moisture content of three grains is shown in table.

Table: Equilibrium moisture contents of three grains (% W.B.).

Relative Humidity	Wheat		Corn		Oll Sunflower	
%	40°F	70°F	40°F	70°F	40°F	70°F
20	8.5	7.7	7.4	6.4	4.6	4.2
30	10.2	9.2	9.3	8.1	5.6	5.0
40	11.7	10.7	11.0	9.7	6.5	5.9
50	13.2	12.0	12.7	11.2	7.4	6.6
60	14.6	13.3	14.5	12.8	8.3	7.4
70	16.2	14.8	16.4	14.5	9.2	8.3
80	18.0	16.5	18.7	16.6	10.3	9.3
90	20.4	18.7	21.7	19.4	11.9	10.7

Drying may take several weeks depending on the airflow rate, climatic conditions and the amount of water to be removed. Natural air/low temperature drying requires enough airflow to complete drying within the allowable storage time. Minimum airflow rates for natural air/low temperature drying of wheat, corn and sunflower are shown in table.

Table: Minimum airflow rates for natural airllow temperature drying of wheat, corn and sunflower.

Maximum Airflow Rate	Inltlal Moisture Content (% WB)		
(cfm/ bu)	Sunflower	Wheat	Corn
½	15	16	18
1	17	18	21
2	21	20	23

A perforated floor is recommended for all in-bin drying. Since air does the drying, it is imperative that air reaches all the grain. Provide one square foot of perforated surface area for each 30 cubic feet per minute (cfm) of airflow. One square foot of bin exhaust opening should be provided for each 1000 cfm of airflow.

The uniform airflow distribution required for drying is more difficult to achieve with ducts than with perforated floors. However, drying can be done successfully with proper duct spacing and careful attention to detail.

Perforated ducts should be placed on the floor with a maximum centerline spacing equal to one-half the grain depth or the shortest distance to the grain surface, and the distance from the duct to the wall must not exceed one-fourth the grain depth at the duct next to the wall. Provide at least one square foot of duct cross-sectional area for each 2000 cfm of airflow. Provide at least one square foot of perforated surface for each 30 cfm of airflow. If the duct is longer than 100 feet, it is better to place a fan at each end of the duct.

Example: A rectangular building 36 by 72 feet is being used to dry wheat. The wheat is spread to a depth of 10 feet (20,800 bushels). At an airflow rate of 1 cfm/bu, a total of 20,800 cfm of air is required. The ducts must not be spaced more than 5 feet apart to be spaced at one-half the grain depth. The distance from the ducts to the wall must not exceed 2.5 feet to be spaced at one-fourth the grain depth. Eight ducts are needed, with the first duct placed 2.5 feet from the wall and the remainder placed 4.5 feet apart. Each duct must handle 2600 cfm of airflow (20,800 ÷ 8). With a velocity of 2000 ft/minute, a duct area of 1.3 square feet is needed (2600 ÷ 2000). This is an 14-inch square duct, a semi-circular duct with a diameter of 25 inches, or a round duct 16 inches in diameter.

The equations to calculate duct cross-sectional area are:

Square or Rectangle

Area (ft²) = Width (in.) x Depth (in.) ÷ 144

Round

Area (ft²) = 3.14 x Diameter (in.) x Diameter

(in.) ÷ 576

Semi-Circle

Area (ft2) = 3.14 x Diameter (in.) x Diameter

(in.) ÷ 1152

Drying 20,300 bushels of wheat using the same airflow rate in the same building with the wheat 6 feet deep on the sidewall and peaked to 15 feet in the center would require nine ducts as shown in figure. The ducts are spaced apart no more than one-half the shortest air path out of the grain. The shortest path is different than the grain depth as shown in the figure. The distance between the wall and the first duct must not exceed one-fourth the grain depth. The duct size varies because the quantity of grain that receives air from the duct varies.

The addition of supplemental heat to the air decreases the final moisture content of the grain. The airflow rate affects the drying rate. Using the temperature and relative humidity of the air after it has been heated, the grain equilibrium moisture content can be determined from table. Heating the air 10 F will reduce the relative humidity about one-fourth, and heating the air 5 F will reduce the relative humidity about one-eighth. With air at 40 F and 80 percent relative humidity, heating it 10 F will reduce the relative humidity to about 60 percent. Grain harvested at or near freezing temperatures may be held over winter at acceptable natural air/low temperature drying moisture contents and dried in the spring. The grain should be cooled to about 25 F for storage during the winter and monitored regularly. Start drying corn and sunflower in the spring as soon as daily temperatures average above freezing (April) and wheat about May 1.

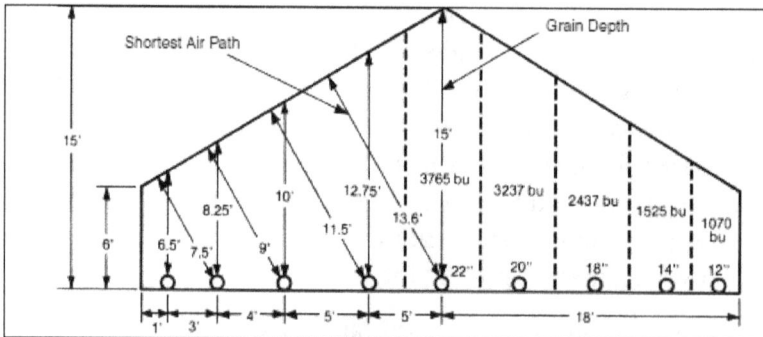

Duct size and spacing for natural air drying in a 36' x 72' building with grain 6 feet deept next to the walls and 15 deep in the center. Perforated duct diameter varies due to different amounts of air required.

The greatest risk involved with natural air/low temperature drying occurs if an abnormally warm, damp period of weather occurs after the grain has been placed in the drying bin. This permits rapid mold development while drying speed is increased very little.

Layer drying or combination drying, are options used with natural air/low temperature drying when the grain is wetter than the system is designed to handle.

Layer Drying

Advantage

- Grain with a higher initial moisture content can be harvested as compared to the maximum initial moisture content used in full-bin drying.

Disadvantage

- The harvesting schedule may be restricted.

Layer drying is very similar to natural air/low temperature drying except the grain is placed into the drying bin in layers normally about 4 to 5 feet deep. An initial batch or layer of grain is placed in the bin and drying is begun. A drying zone is established and begins to move through the grain. Other layers of grain are periodically added so that a depth of wet grain exists ahead of the drying zone. Limiting grain depth to get a higher airflow rate allows drying a crop at higher moisture contents than the system can handle on a full-bin basis. In a bin designed for 1 cfm/bushel on a full-bin basis, the air flow rate is estimated to be about 4 cfm/bushel if the bin is one-fourth full, Figure. The actual airflow rate will vary due to individual fan performance.

The drying front may be found by probing and measuring the moisture content at various levels. Several points should be checked, since progress of the drying front will not be uniform throughout the bin because of fines accumulation. A common problem with layered drying systems is adding additional wet grain too rapidly, resulting in spoilage of the upper layers.

High Temperature Bin Drying

Advantages

- The bin can be used for storage at the end of the drying season.
- Wetter grain can be dried than can be dried with a natural air or low temperature dryer.

Disadvantages

- A large moisture variation between grain kernels is possible.
- Grain damage may occur from stirring.

Batch-in-bin Drying

The batch-in-bin drying process involves using a bin as a batch dryer. A 3 to 4-foot deep layer of grain is placed in the bin and the fan and heater started. Typical drying air temperatures are 120 to 160 F with airflow rates of 8 to 15 cfm/bushel. Dry-ing begins at the floor and progresses upward. Grain at the floor of the bin becomes excessively dry while the top layer of the batch remains fairly wet. The grain is cooled in the bin after it is dried. Some batch-in-bin dryers hold the grain being dried in a layer near the roof. After the grain is dried it is dropped to the bin floor where it is cooled. As the grain is moved from the bin, the grain is mixed, and the average moisture content going into final storage should be low enough that mold growth will not be a problem.

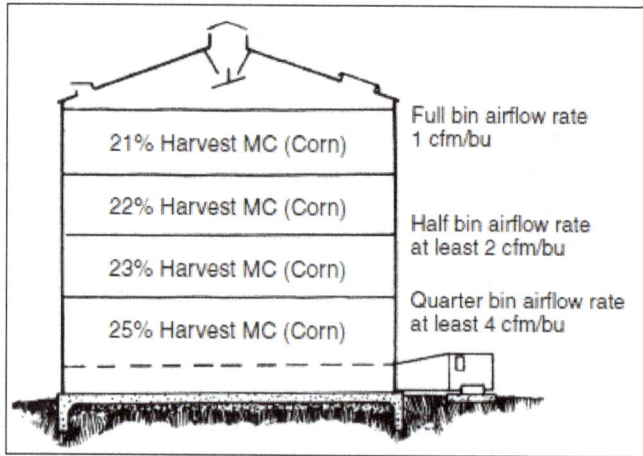

Example of layer drying. The higher airflow rates on a per bushel basis early
in the filling permit a higher initial moisture content to be loaded.

A stirring device can be added to provide more uniform drying and moisture content and to increase the capacity of the bin dryer. Research conducted at Iowa State Univer-sity indicates that with a stirring device there is less than 1 percentage point moisture variation between upper and lower layers of a batch of grain. This research also indi-cates there is some reduction in resistance to airflow, permitting an increase batch size in the typical bin. Stirring allows depths of up to 7 or 8 feet for corn. There is a tendency for fine materials to migrate to the bin floor as the stirring device is in operation.

Condensation is likely to form on the bin walls. If the last batch of grain to be dried is to be left in the bin for the winter months, air tubes and bin liners have been used to help reduce the problems of mold growth next to the bin wall. Another technique that has shown some benefit is to operate the stirring device next to the wall to provide extra stirring.

A disadvantage of batch-in-bin drying is that additional storage for wet grain holding is required.

Recirculating Bin Dryer

The recirculating bin dryer incorporates a tapered sweep auger which removes grain from the bottom of the bin as it dries. The sweep auger may be controlled by tem-perature or moisture sensors. When the desired condition is reached, sensors start the sweep auger, which removes a layer of grain. After one complete revolution around the bin, the sweep auger stops until the sensor determines that another layer is dry. This dried grain is redistributed on top of the grain surface. The dried grain will be partially rewet by the moist air coming through the grain, which reduces drying efficiency. After all the grain has been dried, the grain is cooled in the bin. The dried and cooled grain is then moved to storage or may be left in the bin. It is common to dry the last bin full of grain using a continuous flow bin dryer as a recirculating bin dryer.

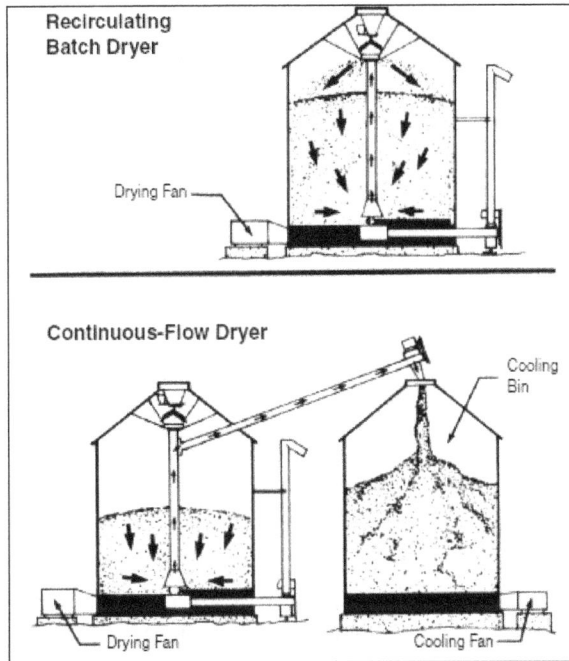

Grain recirculators convert a bin dryer to a high speed recirculating batch or continuous flow dryer.

Continuous Flow Bin Dryer

The continuous flow bin dryer also incorporates a tapered sweep auger which removes grain from the bottom of the bin as it dries, but the grain is moved to a second bin for cooling. Up to 2 points of moisture may be removed in the cooling bin if dryeration is used. Increasing the grain depth will reduce the airflow rate, cfm, and the drying rate of a continuous flow bin dryer. In a recirculating batch or continuous flow bin dryer, it is the total airflow capacity, cfm, that determines the drying rate, not the airflow rate, cfm/bu.

Column Dryers

Advantages

- Dryer does not occupy grain storage space.

- Portable units can be moved from one location to another.

Disadvantage

- The heat available in the dryer is not used as efficiently as in deep bed drying.

Column Batch Dryers

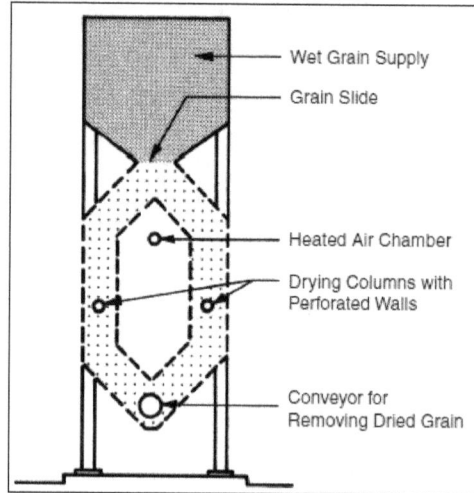

Cross-section of a column batch dryer.

Column batch dryers are completely filled at one time. A common batch dryer configuration is two columns surrounding a plenum chamber. Several circular-shaped batch dryers are also available. Hot air forced into the plenum from a fan-heater unit passes through the grain-filled columns and dries the grain. Common batch capacity of batch dryers varies from 80 to 1,000 bushels. Column widths are normally from 10 to 20 inches. High temperatures and high airflow rates characterize batch dryers. The typical operating sequence is fill-dry-cool-unload. Time for one batch varies, but an average may be two to three hours per batch. Control of the drying sequence can be either manual or automatic.

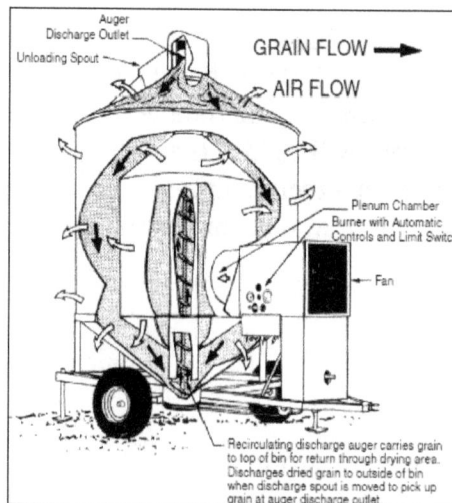

Recirculating batch dryer.

A recirculating device may be added to some batch dryers. This has the effect of reducing the moisture variation across the column of the dryer. For some crops, a higher temperature may be used with a recirculating batch dryer since a kernel of grain will not be next to the heated air for the entire drying cycle and as a result should not get as hot.

Continuous Flow Drying

Wet grain constantly feeds in the top and is dried and cooled in a continuous flow dryer. Dry grain is drawn off the bottom and placed into storage. These dryers are similar to batch dryers in configuration but have a divided plenum chamber. Hot drying air is pushed into the top chamber, and unheated air for cooling is pushed into the lower chamber. Column widths on continuous flow dryers vary from 8 to 20 inches. A sensor controls the discharge rate and consequently the moisture content of the dried grain. Continuous flow dryers use high temperatures and high airflow rates. Airflow rates of 50 to 100 cfm/bushel of grain are common. Continuous flow dryers are available in a large range of sizes. Portable units are available in sizes up to about 1000 bushel per hour capacity, and stationary units of larger capacity are available. The first grain through a continuous-flow dryer generally will need to be cycled through the dryer again for drying to be completed. A continuous flow dryer with cross-flow airflow is shown in Figure.

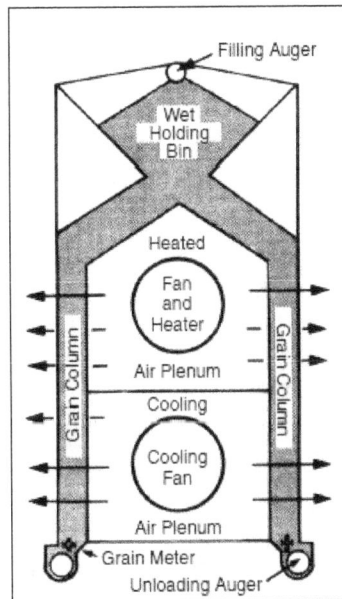

Cross-flow dryer with forced-air drying and cooling.

Some cross-flow models reverse the airflow through the dryer as the grain progresses down the column to reduce overdrying. Some reverse the air flow in the cooling section to increase energy efficiency. A concurrent flow dryer with counter-flow cooling is shown in figure. The concurrent-airflow in the drying section and counterflow in the cooling section improves energy efficiency and reduces stress cracking in corn. With this system, the heated air enters the grain near the top of the dryer and moves downward in the same direction as the grain. The cooling air moves in the opposite direction as the grain.

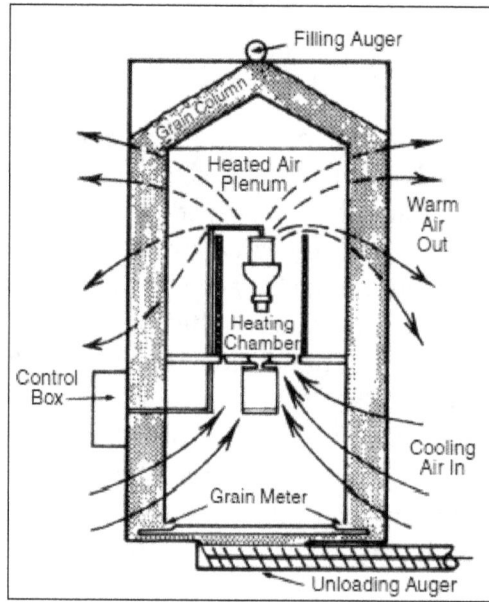

Cross-flow dryer with reverse-flow cooling.

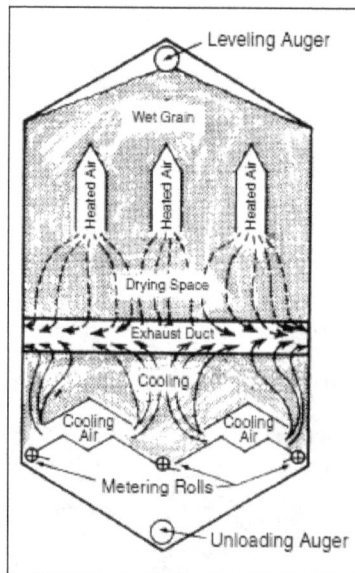

Schematic of a concurrent-flow dryer with counter-flow cooling.

Another type of dryer is the mixed flow dryer shown in Figure. In this type, the grain flows over alternating rows of heated air supply ducts and air exhaust ducts. This action provides mixing of the grain and alternate exposure to drying air that is relatively hot and air which has been cooled by previous contact with the grain. It promotes moisture uniformity and nearly equal exposure of the grain to the drying air.

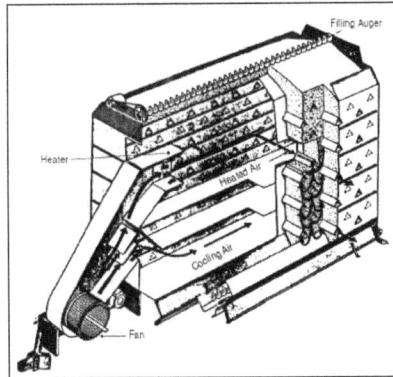

Combination Drying

Advantages

- Increases drying rate of high temperature dryer by about 300 percent.
- Increases energy efficiency.

Disadvantages

- Requires natural air/low temperature drying to complete drying.
- Requires more grain handling.

Combination drying is a process using a high temperature dryer to dry the crop to a certain level, then a natural air/low temperature drying system completes the drying process. This system may be used to increase the capacity of the high temperature drying equipment, for increased energy efficiency, or when conditions are not suitable to start drying with a natural air/low temperature drying system. The crop is dried from the harvest moisture content to a level acceptable for natural air/low temperature drying, then it is moved to the natural air/low temperature dryer and drying is completed.

Dryeration and In-storage Cooling

Advantages

- Increases drying rate by about 60 percent for dryeration and 30 percent for in-storage cooling.
- Increases energy efficiency.

Disadvantages

- Requires cooling fan and bin.
- Requires more grain handling.

Dryeration is a process where hot grain is removed from the dryer with a moisture content 1 or 2 percentage points above that desired for storage. The hot grain is placed in a dryeration bin where it is allowed to temper without airflow for at least four to six hours. The moisture content equalizes in the kernel during tempering. After the first hot grain delivered to the bin has tempered, the cooling fan is turned on while additional hot grain is delivered to the bin. The grain is cooled and 1 to 2 percent moisture content is removed by the airflow before it is moved to final storage. Cooling is normally completed about six hours after the last hot grain is added if the cooling rate equals the filling rate.

In-storage cooling eliminates tempering. Grain is dried to the desired moisture content for storage in the dryer, then moved to storage where it is cooled. Quality of the grain is improved with both in-storage cooling and dryeration because the final drying and cooling are done at a slower rate than in a conventional high temperature drying system.

Haulm Toppers

A Haulm Topper is a device that can prepare Root Crops (Sugar Beet or Potato) for harvesting. These crops cannot be harvested until processed by a Haulm Topper. The Haulm Topper cuts the parts of the plant that are above ground so that a harvester can get at the roots below, which are the actual fruit of these crops.

The game provides separate Haulm Topping machinery for Sugar Beets and for Potatoes. Each crop has one basic, tractor-towed Haulm Topper available; and one combination Haulm Topper/Harvester that can perform the entire harvest in one go.

Trailers

These Haulm Toppers must be towed by a Tractor. They can only perform the Haulm Topping task.

Grimme FT 300

Crop: Sugar Beet

Working Width (m): 3.0

Req. Power*(kW / hp): 45 / 61

Grimme KS 75-4

Crop: Potato

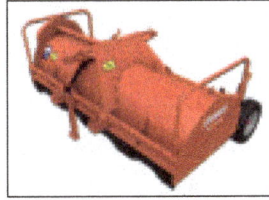

Working Width (m): 3.3

Req. Power*(kW / hp): 45 / 61

Self-propelled

These Haulm Toppers move under their own power. They are actually Harvesters that also top the haulms simultaneously as they work.

Grimme Maxtron 620

Crop: Sugar Beet

Capacity (liters): 32,850

Working Width (m): 3.0

Grimme Tectron 415

Crop: Potato

Capacity (liters): 20,000

Working Width (m): 3.0

The two root crops, Sugar Beet and Potato, are the most profitable direct-sale crops due to the high yields given by each hectare of these crops (compared to all other crops). However, both Sugar Beet and Potato are somewhat more work-intensive than other crops, because of an additional step required to prepare the field for harvesting - the Haulm Topping step.

In essence, once the crops on the field have ripened, a special machine must make a pass over the field to remove the leafy parts of the plant that are sticking above-ground - the "Haulms". This special machine itself is called a Haulm Topper. Each Haulm Topper is designed to work only on one specific type of Root Crop.

The default Haulm Toppers are simple, cheap machines that must be dragged behind a Tractor in order to operate. As the machine passes over the field, it will cut the Haulms off the plants, leaving only the roots underground. These roots, being the actual Beets and Potatoes you need, can then be collected using a separate Harvester device. Until the Haulms are removed, the harvester cannot process the field at all.

The basic Haulm Toppers are easy to use, and work pretty much the same as Cultivators. You may hire a Worker to do the work for you. Note however that the Haulm Toppers available in the base game are quite narrow, and will take some time to process the entire field. There are no wider tools available for this task.

Self-propelled Combination Toppers

In addition to the tractor-pulled Haulm Toppers, the game also offers two Combine Toppers that can move under their own power, not requiring a tractor. In addition to this advantage, the two Combine Toppers are also Harvesters, which will pull the Sugar Beets/Potatoes out of the ground immediately after cutting the haulms. That is, they do both parts of the harvesting process for these crops simultaneously, saving a lot of time and effort.

On the other hand, both of the Combine Toppers in the base game are prohibitively expensive - each costing close to half a million dollars. Of course, if you can gather the money to purchase one, it will likely pay for itself very rapidly, since the Root Crops are quite profitable.

Mechanical Tree Shaker

Mechanical harvesting methods have been investigated and practiced since early 1960s. Coppock stated that citrus fruit can be harvested by mechanical means after proving that a tree could be mechanically shaken to remove the fruit from the branches without destroying the whole tree. To reduce the physical damage to the tree, a pre-harvest abscission spray was also proposed to loosen the fruits on the tree. In order to improve the design of mechanical harvester the biological and physical properties of the fruit were also studied by Coppock. The mechanical harvesting methods reviewed here are limb shaking, air blasting, canopy shaking, trunk shaking, and the use of an abscission chemical agent to loosen the fruits.

Limb Shaker

An early limb shaker was represented by Coppock and Jutras using inertia developed by Adrain and Fridley. An eccentric weight about 85 pounds was rotated in the mechanism to produce the shacking action after the shaker was attached to the tree limb. Notably some damage was made to the bark of the tree by the clamping mechanism. An alternative tree shaker was represented using fixed stoke, inertia, and direct impact on trees limbs. The issues from this practice included such as fruit damage due to fall foliage, lower removal rate in earlier and mid of harvesting season, and large or small immature fruit removal. Another tree shaker with two catching frames each with an inertia type limb shaker was developed. Still, immature fruits were removed with damages to the fruits. A self-propelled limb shaker was tested on Valencia orange. A self-propelling full powered positioning limb shaker was also evaluated with abscission aid by Summer and Hedden.

Air Blast

The application of force generated by airblast to remove the fruit started in 1961. An oscillating air blast machine was tested and practiced by Jutras and Patterson. Fruit removal was maximized by the oscillation rate. The air blast model and all the subsequence models were made and named after FMC (Food Machinery Corporation. The performance of FMC series was dependent on factors such as structure of tree, size and weight of fruits. Later, an air shaker was designed and constructed to alleviate issues such as the high power requirement. However still damages to the fruits and leaves were the major issues addressed in the project.

Canopy Shaker

A canopy shaker was designed to clamp secondary limbs to shake vertically. The shaker was extended into tree with a pantograph lift unit and shake always vertically. An excessive immature oranges were removed during tests conducted by Summer. Two continuing canopy shakers were reported by Futch and, one was self-propelled unit and another was tractor-drawn unit. These two units were used for juice processing plants. Manual workers were needed to collect the fruits after the harvest. Shaking frequency and stroke are important factors in the performance in this type of harvester and it requires more tests to determine the optimal values.

Trunk Shaker

Trunk shaker was used to remove deciduous fruits and nuts but difficult to apply this technique on citrus fruit removal. However, Hedden and Whitney designed the experiment to evaluate the trunk shaker for earlier season Hamlin orange and late season Valencia orange using different unbalanced mass and multidirectional shakers for years. The linear low frequency shaker with a larger displacement performed better than the canopy shaker machine. The bark was more or less damaged during the experiment. Later the trunk shakers were tested along with other canopy shakers by Whitney. The efficiency or removal was from 67% on large trees to 98% on small trees. More recently, a tractor mounted trunk shaker was tested on varieties of oranges and mandarins in Spain by Torregrosa in comparison to a hand-held shaker. Overall the tractor mounted shaker was more effective with 72% detachment than the hand-held shaker with 57% detachment. In test, the fruits picked up from ground had high percentage of bruise. Defoliation was high at high shaking frequency and the bark was damaged in season of May and June.

Abscission Chemical

Abscission chemical agent was designed to loosen the mature fruit and improve the rate of removal of fruit in harvesting season. There are many kinds of abscission agent such as Ethephon and 2-chloroethyl phosphoric acid. The use of abscission agent was applied as pre-harvest process and constituted part of harvesting such as air. Air shaker was tested with applying abscission agent in advance on FMC-3 by Wilson. Limb shaker used abscission to loosen the fruit on stem. It was noted that abscission chemical was inconsistent in practical use. The abscission use was subjective to many factors such as weather factors, tree injury, and cost of using chemical such as equipment. The Prosulfuron, an abscission chemical agent which was used on Hamlin and Valencia oranges loosening, were studied by. This abscission was more effective in Hamlin than others. However, the immature Valencia was loosen before harvest. The CMN-P abscission chemical was tested on 'Hamlin' orange before the harvesting by trunk shaker. Three abscission chemicals and four types of trunk shaker with scissors type clamps were used in the test. Usually the chemicals helped to loosen both Hamlin and Valencia oranges and improved the removal rate by 30%. The main issue from the test was the pre-harvest drop of fruits. The chemical Methyl Jasmonate (MJ) was tested on whole tree or canopy sectors of Valencia oranges. High MJ caused great loosening of fruit but excessive defoliation as well. The combined use with other compounds was proposed for later experiment especially on Valencia orange harvest. A CMNP abscission chemical was tested on Hamlin and Valencia oranges by Kostenyuk and before the mechanically harvesting. For all of these methods mentioned, the consistency of the performance is the major issue.

Post-harvest

Peripheral mechanical harvesting systems such as catching frames have been designed to reduce the damage on falling fruits. These frames also help to collect the fruit efficiently. In addition, a tractor pulled rake was made to windrow the fruits to the drip-line and the windrowed machine was used to pick up windrowed fruits. Both methods were not adaptable to all grove conditions. The grove conditions should be considered to maximize the efficiency of all mechanical harvesters based on the comparison study on different mechanical harvesters. The quality of fruits picked up by a harvesting machine was evaluated. Also the potential for microbiological contamination to the process was also presented by Goodrich et al.

Automatic Harvester

The mechanical harvesting system cannot maintain the quality and size selection that the human vision can. The automatic individual harvester was considered as an alternative method to the mechanical harvester by Schertz and Brown. The two detachment devices, a vacuum twist device and a rotating cutoff device, were used in the research. The photometric comparisons showed the potential use of the light reflectance for the fruit detection. The concepts were further developed by Parrish and Goskel from University of Virgia in 1976. Much concrete works began at around 1983 at Kyoto University at Japan, and at University of Florida. Then CEMAGREF Montpellier extended it on the subsequent projects. The study focused on the sensors and the vision system in mechanical manipulator. Even though the review is majorly focused on the citrus fruit harvester, some other significant robotic harvesting systems for different kinds of fruits harvesting are selected and reviewed as a complementary reference.

Magali Project

The basic idea of automatic harvesting system was early developed at Montpellier France. A camera was fixed in the pathway to detect the fruits and a grasping tool was sent to the fruit within straight line trajectory. This basic concept was tested on a labouratory prototype in May 1984. A simple black and white(B/W) camera was positioned above the telescope arm aiming fruit horizontally in this model. Years later, the video colour camera was applied to replace the B/W camera with a proximity sensor to sense the fruit touch in grasper. This first prototype of self-propelled robot was built to harvest the golden apple. The prototype used three degrees of freedom and hydraulically powered spherical coordinate with joint position compensation and an end effecter shaped as suction cup. The camera was positioned at the centre of rotary joints at the base of the manipulator. The vision coordinates coincided with the robot coordinates making the end effecter in "line of sight". More than 50% fruits were detected and harvested with 75% fruits with the stem on. In average, each fruit was harvested in around 4 seconds.

Florida Citrus Picking Robot

Issue for the "line of sight" from MAGALI project was the obscure between the vision and the target during the later picking stage. Inaccuracy of guidance was accounted when the fruit was moved by wind. Harrel et al overcame this issue by attaching the colour video at the end of the effecter on a spherical manipulator with an ultrasonic transducer to compensate the distance to the fruit. This "eye-inhand" approach used the contrast colour between the fruit and the background to track the fruit. The cycle for harvesting was about 3 seconds to 7 seconds with 75% fruit detected.

Eureka Projects

Spanish and French research team cooperated on a project to harvest citrus fruits. The gripper was a spherical manipulator with hydraulic power, and the camera positioned in the centre of the manipulator. In vision detection, a single B/W camera of gray level scheme was tested using 635nm wavelength filter and 560nm wavelength filter with a supported flashing light. Then the colour image for the mature fruit detection was used with Bayesian classifier as discrimination function. The colour scheme was superior to the monochromatic scheme with over 90% fruits detected. The occlusion by leaves was one issue found in the project. The most failures were the distance problem in the approaching stage when it is close to the fruits. Hence the need of knowing distance from manipulator to the fruit target is of all importance in tests.

Agribot Project

Agribot project from Spain developed a semi-automatic harvester by combining both human and machine functions. Two jointed harvesting arms were built and were mounted on the human guided vehicle. The manipulator was three degree of freedom (DOF) design with one vertical rotation axes and two horizontal rotation axes and the rotation on the end effector with gripper. The gripper was specially designed with the pneumatic suction cup and proximity IR sensor to sense the attached fruit followed by the saw cutter to cut the stem. Detection was done by the human operator when the vehicle was placed opposite the fruit tree using a joystick to control the pan and tilt mechanism pointing to the target. Fruits were localized in spherical coordinates with distance detected by laser telemetry and the angular position by two rotational axes. The laser ranger finder was sensitive to the outdoor lighting conditions. Hence the special goggles were used to block and enhance the red laser spot. The harvesting cycle time for this machine was 2 second.

Cram Citrus Picking Robot

In Italy, researchers and collabourators from the University of Catania developed a citrus picking robot. A Cartesian based manipulator was housed on a vehicle with a caterpillar to maximize the reachability. The top arm was equipped with a pneumatic piston to extend the gripper far inside the canopy while the lower arm was not equipped the pneumatic extension. In the vision system, the index RG from RGB colour model was adopted to segment fruits from the background. The fruit was located by estimating the diameter. The distance was estimated based on the dynamic incremental movement of manipulator. Kalman filter was used to estimate the real dimension of the fruit in the reaching stage in sequential estimation. In simulation test, the picking time per fruit is about 5.93 second.

Other Fruit and Vegetable Robotic Harvester

Some other robotic harvesters have been reviewed and listed. from Italy presented a robotic harvesting system on AGROBOT project to harvest the tomato in green house. The system was based on a vehicle carrying a six DOF of picking arm consisting of gripper and hand, two micro cameras, and the control system with the image frame grabber and the image processer. The components of hue and saturation from HSI colour space were used to perform threshold to segment the image. A prototype of robotic melon harvester was developed and practiced by joint researchers from Israel and. In Japan, has investigated harvesting method with a robot for tomato, cucumber, and grape. In Netherland, presented a greenhouse robot to harvest the cucumber using an industrial manipulator with 7DOF to position the end effecter. Hayashi presented an eggplants robotic harvester. In Belgium, integrated the industrial manipulator with the newly designed and patented flexible gripper which consisted of the silicon funnel with camera mounted inside the center of the funnel to harvest apples.

Machine Vision System in Automatic Harvesting

The vision system used in the automatic harvester aims to detect the fruits and provides the information of the location and the distance to the fruits to the robotic controller. In vision recognition system, vision cameras are mainly the solution to communicate with the environment. The major achievements of vision systems and the performance of the various sensors in the harvesting have been reviewed by Jimenez. The sensor techniques applied in agricultural applications have also been reviewed by Chen. Normally, the digital image data has been categorized in three types; intensity, spectral, and laser range finder. On top of the basic colour processing, new technologies have also been investigated especially in the measurement and spectral analysis such as the hyperspectral imaging technology(HIS) and the thermal imaging technology. The visibility of the fruit citrus on canopy has been studied using multiple perspective viewing method analysis as well for a harvesting robot.

Monocular Scheme

In applications, a single gray scale camera or a colour camera has been widely applied on citrus or orange detection. The Italian AID robot vision system adopted a colour camera with artificial lighting to detect oranges. Approximately 70% of the visible fruit were recognized. Slaughter and Harrell used a digital colour camera with a filter of 675nm wavelength to enhance the contrast between orange fruits and others. Over 75% orange were detected with some non-fruit part misclassified. There are three separate vision system development projects found in European Eureka Project. A single grayscale camera was tested using 635nm filter for citrus and 560nm filter for leaves. Then two cameras with a red filter and a green filter were tested to segment the fruit respectively. Finally a colour image for mature fruits was used for fruit identification with over 90% fruits detected. Regunathan and Lee used a colour camera to capture the citrus fruit with an ultrasonic sensor to obtain the distance between fruit and camera. Issues such as partial occlusion, variable illumination, and clustering of fruit caused errors in both fruit count and size estimation were addressed.

A single camera has been applied also in apple detection. used a single gray scale camera with a colour filter attached to detect the fruit apples. In MAGALI project, both B/W and colour camera were used to detected the. Later to overcome the obscure issue of this scheme, Harrell designed the vision system by positioning the colour video at the end of effecter on a spherical manipulator with an ultrasonic transducer to compensate the distance to the fruit. The success rate was about 75% for picking attempts and about 50% from all fruits. Sites used a single CCD camera to catch gray level fruit images of peach and apple with filters. Up to 92% fruits were detected. Bulanon used a colour CCD camera to acquire images and 80% apples were detected under natural lighting condition. Zhao used a single colour image to segment the fruit apples using both colour and texture. In tests, 18 apples were recognized out of 20 visible apples. One of interesting practices with video sequences was presented by Tabb to segment the apples fruit in labouratory environment. Up to 95% of apples were identified. In Belgium, Baeten used the colour camera positioned inside the center of the gripper in a funnel shape. About 80% apples were detected and harvested in the experiment.

Some other fruits and vegetables have also been practiced using this scheme. used a greyscale video camera to capture the tomato images. Average 68% tomato was detected with a proper threshold. In melon harvesting system, two single B/W cameras were positioned as far vision and near vision sensors for planning and controlling level respectively. About 80% melons were detected in experiment. The author proposed the improvement by combining infrared and visible image or by merging multiple sensor fusion methods. In AGROBOT project, two colour cameras were positioned on each of two DOF manipulator head mechanism. Later it was also found that the use of stereoscope matching scheme could extract the 3D information of object from both segmented results from two colour images.

Binocular Stereoscope Scheme

An early approach using stereo vision system was presented by Kassay from Hungary AUFO robot project. Two colour cameras were designed and positioned separately in a distance with an angle converging to the same scene. The photogrammetric principle was applied based on a triangulation algorithm. In the experiments about 41% of visible apples were detected. Similar to this scheme, developed a stereo matching algorithm to recognize oranges using two images captured from different view of angles. Another stereo matching method was presented by Plebe and in which two cameras were carried in the center of the gripper on each of two robotic arms. The 3D match was done based on disparity constraints. In Italian AGROBOT project, two colour cameras was designed and positioned on two DOF manipulator. The 3-dimensional information was then extracted from stereoscope matching of two tomato images on the same scene. Takahashi et al presented another binocular stereo vision system. Some issues were found commonly. The measured disparity was difficult to guarantee the estimation of the dynamic shape of the natural fruits. The intensive computation of transforming the disparity between the objects in two images into depth information was costly. The lighting conditions and occlusion problems affected the detection efficiency.

Structured Light Scheme

Jimenez et al presented a laser range finder sensor for AGRIBOT robot on orange detection. This sensor gave the spherical coordinates of the scanned scene points. The range and attenuation data values of scanned surface were identified by the laser energy depending on the distance, surface type, and the orientation of the sensed points. Totally four types of images were used in the system. In the image processing, the shape analysis could detect both mature and green fruits if the contour of the fruit was visual more than 70% of the whole. Overall 87% visible fruits were detected without any false detection. Benady and presented the use of the structure light with a laser profiling system to locate the melon. Saint-Marc placed two cameras on both sides of the sheet light plane with an angle to overcome the occlusion problem. Two scanned images from both cameras were matched based on the calibration. In Japan, a custom built laser range finder was applied by for tomato, cucumber, and grape harvesting. The custom-built active range finder projected and received two lasers light in the same optical axis and the distance from the sensor to the target was calculated by the demodulated signal from both beams. A similar idea was found in a cherry robotic harvesting system from Japan by. The 3D sensor was custom built and equipped with red and infrared laser diodes to detect the ripe cherry. In the experiment, 10 out of 12 fruits were recognized. The use of structured light had faced some issues such as the sparseness of measurements, occlusion, and the reading of the orientation of the surface. Scanning could be improved by parallelizing the scanning and the analysis process.

Hyperspectral Scheme

Hyperspectral technique has been applied in quality inspection analysis and measurement applications. Generally, the hyperspectral sensor is designed to disperse the whole sensible spectral area into discrete wavebands of spectra. In the output image, each pixel collects stacks of the waveband segments continually across the whole sensible spectrum. A hyperspectral system was developed by to detect the defects and contaminations on apples. The wavebands between 542nm to 752nm gave a great statistical disparity between defect surface and normal surface of apples. used an airborne hyperspectral remote sensing to predict the citrus yield. Ariana et al used a near infrared hyperspectral imaging to detect the bruises on pickling cucumbers. In the experiment, the waveband ratio and difference were preferable since the classification between the normal and defection areas was more consistent in those two. Okamoto present a green citrus fruit detection using a hyperspectral imaging camera. The camera employed could sense from 369nm to 1042nm waveband areas. Three different types of green citrus fruits were captured by this sensor; Tangelo, Valencia, and Hamlin. The time to acquire a hyperspectral image varied from 22s to 65s depending on the scanning area and the image resolution. The main cause of incorrect identification was the similar spectral characteristics between the green leaves and the green fruits.

Thermal Imaging Scheme

The infrared thermograph has been applied in application using sensors responding to longer wavelength. Stajnko developed a method for estimating the number and the diameter of apples in an orchard using a thermal camera. The thermal sensing was based on the heat emission depending on the exposure time from the sun heat. A recent project using thermal infrared imaging in fruit citrus canopy was presented by. The thermal radiation which was based on the Stephan-Boltzman law was used as an indicator of the temperature of the emitting object. It was found that the large temperature difference between the fruit and the canopy occurred around 4:00 pm. The histogram based algorithm performed well during the selected time. Later the author developed a fusion technique to improve the fruit detection using both colour image and thermal image. The registration was required to align two different images from thermal sensor and the visible camera by a half-meter PVC pipe frame. In the experiment, the success rate was high when the temperature of fruit was warmer than the canopy. The fusion technique using thermal imaging and colour imaging could be a practical alternative during an appropriate time period.

Multispectral Scheme

Instead of dispersing spectrum into the discrete numerous wavebands, multispectral image capture the specific across wavelength area in the spectrum. The early multispectral scheme was presented. The custom built camera consisted of three CCD micro cameras side by side with three different optical waveband filters in terms of 550nm, 650nm, and 950nm wavelength respectively. Three images were calculated and combined to generate a binary image. The detection of fruit was about 75% when the sky was overcast. This scheme initially gave the implication to possibly detect immature green fruits by combining different waveband spectra. Kane and used a monochromatic near infrared camera equipped with multi waveband pass filters to capture citrus fruit tree images. Averagely 84.5% correct citrus pixels were identified. This work was extended after the measurements on the green leaves and the green citrus fruits through types of citrus and seasons. Three waveband filters were attached to the camera respectively to catch waveband spectral area images. However the resultant multispectral images were not well synchronized and matched in dynamic scenes. Another issue was the saturation of sensor value and the dark area on the image. The reason that the number of leaves caused the diffuse reflectance was theoretically studied. It was proposed that the multispectral images should be captured at the same time with capability of acquiring more wavebands and a smarter image processing technique. carried out their preliminary investigation on the detection of melon using infrared image. They proposed the method by combining the infrared and visible image to improve the detection result.

The multispectral scheme has been more designed in inspection analysis applications. In Spain, presented a machine vision scheme on citrus online inspection using two CCD cameras with one camera for visible area detection and the second for infrared area detection. Lu et al presented a multispectral scheme to predict the firmness and the soluble solids content of the fruit apples. More recently, has developed two types of portable multispectral imaging system, a dual-band spectral imaging system and three-band spectral imaging system for contaminant detection in food inspection industry.

Post-harvest Automation

Post-harvest technology is inter-disciplinary "Science and Technique" applied to agricultural produce after harvest for its protection, conservation, processing, packaging, distribution, marketing, and utilization to meet the food and nutritional requirements of the people in relation to their needs. It has to develop in consonance with the needs of each society to stimulate agricultural production; prevent post-harvest losses, improve nutrition and add value to the products. In this process, it must be able to generate employment, reduce poverty and stimulate growth of other related economic sectors. The process of developing of post-harvest technology and its purposeful use needs an inter-disciplinary and multi-dimensional approach, which must include, scientific creativity, technological innovations, commercial entrepreneurship and institutions capable of inter-disciplinary research and development all of which must respond in an integrated manner to the developmental needs.

Importance

Importance of post-harvest technology lies in the fact that it has capability to meet food requirement of growing population by eliminating avoidable losses making more nutritive food items from low grade raw commodity by proper processing and fortification, diverting portion of food material being fed to cattle by way of processing and fortifying low grade food and organic wastes and by-products into nutritive animal feed. Post-harvest technology has potential to create rural industries. In India, where 80 percent of people live in the villages and 70 percent depend on agriculture have experienced that the process of industrialization has shifted the food, feed and fibre industries to urban areas. This process has resulted in capital drain from rural to urban areas, decreased employment opportunities in the rural areas, balance of trade in favour of urban sector and mismatched growth in economy and standard of living including the gap between rural and urban people. It is possible to evolve appropriate technologies, which can establish agricultural based rural industries.

The purpose of post-harvest processing is to maintain or enhance quality of the products and make it readily marketable. Prime example of post-harvest processing of agricultural products is rice, a major crop in India. Paddy is harvested and processed into rice. Experiments with paddy crop in farmer's field in India have shown that if the crop is harvested at 20 to 22 per cent moisture as traditionally done, the field yield is increased by 10 to 20 percent. Similar is the case with respect to wheat, jowar and other crops.

Post-harvest Losses

Due to old and outdated method of paddy milling, improper and inefficient methods of storage of paddy, rice, transport and handling we lose about nine percent of production. The traditional methods of storage are responsible for about six percent losses. If better methods of processing and storage are adopted, the losses could be reduced to 2 to 3 percent and more food grains could be available to the people. It is estimated that 10-15 percent of horticultural crop such as vegetables and fruits perish due to lack of proper methods of processing and storing. The loss in monetary term is estimated to be about Rs.20 crores annually.

Proper methods of processing, storage, packaging, transport and marketing are required for export of crops such as jute, tea, cashew nuts, tobacco, mango, litchi, nut, spices and condiments. One of the attributes to this post-harvest system, as it is now constituted, is the large amount of wastage it involves. In case of food grains, some estimates suggest that in developing countries as much as 1/4th to 1/3rd of total crop may be lost as a result of inefficiencies in the post-harvest system. Losses of food crops refer to many different kinds of loss produced by a variety of factors. These include weight loss, loss of food values, loss of economic value, loss of quality or acceptability and actual loss of seeds themselves.

Priorities and Strategies

The priority areas in food processing are:

- Processing of special fruits and nuts like, banana, litchi, mango, pineapple, makhana etc. and canning and storage facility for the above produce.

- Large scale introduction of mini rice mill in villages and mandies coupled with semi-modern parboiling plant for paddy to have higher head rice recovery with better quality bran. Oil production from bran with a chain of collection mechanism for supplying raw material for the plant.

- More emphasis on the use of power Ghani or expeller in place of Kolhu for higher recovery.

- Establishment of dal mills in pulse growing belt as a village cooperative programme.

- Emphasis on cottage industry involving village women for the manufacture of food products.

- Popularization of low cost engineering storage structures.

- Starch production from maize and potato and simultaneous oil production from maize.

- Strengthening of research base with adequate financial support.

- Emphasis on production of value added products from locally available fruits and vegetables.

There are many other areas of processing aspects which should be given priority. Once the processing, research and industry programme picks up many other outlets shall come up automatically. Processing industry has very good employment potential.

Post-harvest Industries

- The post-harvest industry includes the following main components.

- Harvesting and threshing.

- Drying and storage.

- Processing (conservation and / or transformation of the produce).

- Utilization by consumer including home processing.

Other Components of the System

- Transportation and distribution.

- Marketing.

- Grading and quality control.

- Pest control.

- Packaging.

- Communication among all concerned.

- Information, demonstration and advisory systems.

- Manufacture and supply of essential equipment and machinery.

- Financial control.

- Price stabilization

- Management and integration of the total system.

Potential of Income and Employment Generation through Post Harvest Operations

Use of appropriate post-harvest technology reduces the post-harvest and storage losses; adds value to the product, generate employment in village and reestablishes agro-industries in rural sector. Presently, the farmers sell their products without processing. If they do primary processing and value addition in the villages, it will generate more income and employment in rural sector. The processing of food, feed, fibre, oilseeds and sugarcane will generate enough employment in rural areas. If an agro processing center is established in each big village or a cluster of small villages for primary processing, it will generate employment to about 4-5 persons and will increase income of the farmer/processor by about 15-20 percent. Use of proper post-harvest technology of perishables and semi-perishables will reduce the wastage to great extent.

Potential of Income and Employment Generation in other Areas

Besides potential for income and employment generation in crop production and post-harvest sector, there is a great potential of income and employment generation in allied sectors also by using Agricultural Engineering Technology. They are:

- Animal production - Fodder, boiling, briquething, palleting, - Poultry feed Industry.

- Fish Production - Improved hatchery, production and transport of fish seed fingerlings etc.

- Dairying - Processing of milk and making dairy products.

- Energy management in agriculture - Efficient use of biomass, wind and solar energy.

- Wasteland development - Conservation of wasteland, afforestation, management of trees and grasses.

- Agro-forestry.

- Watershed Management.

Action Taken to Tap the Potential

In order to take advantage of agricultural engineering technology for generating income and employment in rural areas following actions are suggested:

- Bringing awareness amongst the rural people about the new developments in agricultural engineering technology in different fields.

- Organizing training programmes for the farmers/agricultural laboures/entrepreneurs about the use of new technology.

- Mass production of different types of agricultural machinery for farmers/entrepreneurs.

- Starting agro - processing centers in each village for primary processing of food grains, fruits and vegetables.

- Providing institutional credit for the purchase of agrl. machinery and starting agro-processing centres.

- Developing market network for purchase/supply of processed material from agro-processing centers.

- Developing proper network and infrastructures for popularization of agricultural machinery for crop production, and setting up agro-processing centers.

Packaging Techniques for Fruits and Vegetables

India is a land of large varieties of fruits and vegetables due to its vast soil and climatic diversity. With 38 and 71 million tons of production of fruits and vegetables, India is the second largest producer of fruits and vegetables next to Brazil and China respectively. It is also a matter of concern that there is a wide gap between availability and the per capita nutritional requirement of fruits. The low availability of quality fruits & vegetables is mainly due to considerably high post-harvest losses, poor transportation facilities, improper storage and low processing capacity coupled with the growing population. Around 20-30% losses take place during harvesting, grading, packaging, transportation and marketing of fruits.

The fruits of increased production of fruits and vegetables and other agricultural produce will be realized only when they reach the consumer in good condition and at a reasonable price. The existing post-harvest loss of fruits and vegetables could be considerably reduced by adopting improved packaging, handling and efficient system of transport. Packaging of fruits and vegetables is undertaken primarily to assemble the produce in convenient units for marketing and distribution.

Requirements of Packaging

The package must stand up to long distance transportation, multiple handling, and the climate changes of different storage places, transport methods and market conditions. In designing fruit packages one should consider both the physiological characteristics of the fruit as well as the whole distribution network.

The package must be capable of:

- Protecting the product from the transport hazards,

- Preventing the microbial and insect damage,

- Minimizing the physiological and biochemical changes and losses in weight.

The present packaging systems for fresh vegetables in our country is unsuitable and unscientific. The uses of traditional forms of packages like bamboo baskets are still prevalent. The other types of packages generally used are wooden boxes and gunnysacks. The use of corrugated fiberboard boxes is limited. The use of baskets besides being unhygienic also does not allow adequate aeration and convenience of easy handling and stocking. Considering the long term needs of eco-systems and to achieve an overall economy, other alternatives available like corrugated fibre board boxes, corrugated polypropylene board boxes, plastic trays/crates/wooden sacks, moulded pulp trays/thermoformed plastic trays and stretched film and shrink wrapping would have to be considered. Modern packages for fresh fruits and vegetables are expected to meet a wide range of requirements, which may be summarized as follows:

- The packages must have sufficient mechanical strength to protect the contents during handling, transport and while sacked.

- The construction material must not contain chemicals, which would transfer to the produce and cause toxic to it or to humans.

- The package must meet handling and marketing in terms of weight, size and shape.

- The packages should allow rapid coding of the contents.

- The security of the package or its ease of opening and closing might be important in some marketing situations.

- The package should identify its contents.

- The package must be required to aid retail presentation.

- The package might need to be designed for ease of disposal, reuse or recycling.

- The cost of the package should be as less as possible.

Packaging may or may not delay or prevent fresh fruits and vegetables from spoiling.

However, incorrect packaging will accelerate spoilage. Packaging should serve to protect against contamination, damage and excess moisture loss.

Packaging Materials in Use

A great variety of materials are used for the packing of perishable commodities. They include wood, bamboo, rigid and foam plastic, solid cardboard and corrugated fibre board. The kind of material or structure adopted depends on the method of perforation, the distance to its destination, the value of the product and the requirement of the market.

CFBC Boxes

Corrugated fiberboard is the most widely used material for fruit & vegetable packages because of the following characteristics:

- Light in weight.

- Reasonably strong.

- Flexibility of shape and size.

- Easy to store and use.

- Good pointing capability.

- Economical.

Wooden Boxes

Materials used for manufacture of wooden boxes include natural wood and industrially manufactured wood based sheet materials.

Sacks

Sacks are traditionally made of jute fibre or similar natural materials. Most jute sacks are provided in a plain weave. For one tonne transportation of vegetables, materials of 250 grams per square meter or less are used. Natural fibre sacks have in many cases been replaced by sacks made of synthetic materials and paper due to cost factors, appearance, mechanical properties and risk of infestation and spreading of insects. Sacks made of polypropylene of type plain weave are extensively used for root vegetables. The most common fabric weight is 70-80 grams per square meter.

Palletisation

Pallets are widely used for the transport of fruit & vegetable packages, in all developed countries.

The advantages of handling packages on pallets are:

- Labour cost in handling is greatly reduced.

- Transport cost is reduced.

- Goods are protected and damage reduced.

- Mechanized handling is very rapid.

- Through high stacking, storage space can be more efficiently used.

- Pallets encourage the introduction of standard package sizes.

Ventilation of Packages

Reduction of moisture loss from the product is a principal requirement of limited per-meability packaging materials. A solution to moisture loss problems from produce appeared with the development and wide distribution of semi permeable plastic films. Airflow through the ventilation holes allows hot fruit or vegetable to slowly cool and avoid the buildup of heat produced by the commodity in respiration. Holes are also important in cooling the fruit when the packages are placed in a cold storage, especially with forced air-cooling. Ventilation holes improve the dispersal of ethylene produced.

Cushioning Materials

The function of cushioning materials is to fix the commodities inside the packages and prevent them from mixing about in relation to each other and the package itself, when there is a vibration or impact. Some cushioning materials can also provide packages with additional stacking strength. The cushioning materials used vary with the commodity and may be made of wrapping papers, fibreboard (single or double wall), moulded paper pulp trays, moulded foam polystyrene trays, moulded plastic trays, foam plastic sheet, plastic bubble pads, fine shredded wood, plastic film liners or bags.

Controlled and Modified Atmospheric Packaging

The normal composition of air is 78% Nitrogen, 21% Oxygen, 0.03% Carbon dioxide and traces of other noble gases. Modified atmosphere packaging is the method for extending the shelf-life of perishable and semi-perishable food products by altering the relative proportions of atmospheric gases that surround the produce. Although the terms Controlled Atmosphere (CA) and Modified Atmosphere (MA) are often used interchangeably a precise difference exists between these two terms.

Controlled Atmosphere

This refers to a storage atmosphere that is different from the normal atmosphere in its composition, wherein the component gases are precisely adjusted to specific concentrations and maintained throughout the storage and distribution of the perishable foods. Controlled atmosphere relies on the continuous measurement of the composition of the storage atmosphere and injection of the appropriate gases or gas mixtures into it, if and when needed. Hence, the system requires sophisticated instruments to monitor the gas levels and is therefore practical only for refrigerated bulk storage or shipment of commodities in large containers.

If the composition of atmosphere in CA system is not closely controlled or if the storage atmosphere is accidentally modified, potential benefit can turn into actual disaster. The degree of susceptibility to injury and the specific symptoms vary, not only between cultivars, but even between growing areas for the same cultivars and between years for a given location. With tomatoes, excessively low O2 or high CO2 prevents proper ripening even after removal of the fruit to air, and CA enhances the danger of chilling injury.

Modified Atmospheric Packaging

Unlike CAPs, there is no means to control precisely the atmospheric components at a specific concentration in MAP once a package has been hermetically sealed.

Modified atmosphere conditions are created inside the packages by the commodity itself and/or by active modification. Commodity generated or passive MA (Modified Atmosphere) is evolved as a consequence of the commodity's respiration. Active modification involves creating a slight vacuum inside the package and replacing it with a desired mixture of gases, so as to establish desired EMA (Equilibrated Modified Atmosphere) quickly composed to a passively generated EMA.

Another active modification technique is the use of carbon dioxide or ethyl absorbers (scavengers) within the package to prevent the build-up of the particular gas within the package. This method is called active packaging. Compounds like hydrated lime, activated charcoal, magnesium oxide are known to absorb carbon dioxide while iron powder is known as a scavenger to carbon dioxide. Potassium permanganate and phenyl methyl silicone can be used to absorb ethylene within the packages. These scavengers can be held in small sachets within the packages or impregnated in the wrappers or into porous materials like vermiculite. For the actively respiring commodities like fruits and vegetables, the package atmosphere should contain oxygen and carbon dioxide at levels optimum to the particular commodity. In general, MA containing between 2-5% oxygen and 3.8% carbon dioxide have shown to extend the shelf life of a wide variety of fruits and vegetables. If the shelf life of a commodity at 20-25°C is one day, then by employing MAP, it will get doubled, whereas refrigeration can extend the shelf life to 3, and refrigeration combined with MAP can increase it to four days. Few types of films are routinely used for MAP. The important ones are polyvinyl chloride, (PVC), polystyrene, (PS), polyethylene (PE) and polypropylene (PP). The recent developments in co-extrusion technology have made it possible to manufacture films with designed transmission rates of oxygen.

Vacuum Packaging

Vacuum packaging offers an extensive barrier against corrosion, oxidation, moisture, drying out, dirt, attraction of dust by electric charge, ultra violet rays and mechanical damages, fungus growth or perishability etc. This technology has commendable relevance for tropical countries with high atmospheric humidity.

In vacuum packaging, the product to be packed is put in a vacuum bag (made of special, hermetic fills) that is then evacuated in a vacuum chamber and then sealed hermetically in order to provide a total barrier against air and moisture. If some of the product cannot bear the atmospheric pressure due to vacuum inside the package then the packages are flushed with inert gases like Nitrogen and CO_2 after evacuation.

Edible Packaging

An edible film or coating is simply defined as a thin continuous layer of edible material formed on, placed on, or between the foods or food components. The package is an integral part of the food, which can be eaten as a part of the whole food product. Selection of material for use in edible packaging is based on its properties to act as barrier to moisture and gases, mechanical strength, physical properties, and resistance to microbial growth. The types of materials used for edible packaging include lipids, proteins and polysaccharides or a combination of any two or all of these. Many lipid compounds, such as animal and vegetable fats, acetoglycerides have been used in the formulation of edible packaging for fresh produces because of their excellent moisture barrier properties. Lipid coatings on fresh fruits and vegetables reduce weight losses due to dehydration during storage by 40-70 per cent. Research and development effort is required to develop edible films and coatings that have good packaging performance besides being economical.

Improved packaging will become more essential in India as International trade expands after globalization. Standardized packaging of sized and graded produce that will protect the quality during marketing can greatly aid transactions between sellers and buyers. Better packaging should be of immediate value in reducing waste. Much background research on packaging of perishable products and flowers is needed simulating the actual handling conditions expected during marketing.

Value Addition

After harvest the biological produce can be either preserved or processed. Value addition is a terminology used to define the processing of biological produce. Through processing the value of the commodities can be increased by converting it to different products by using conventional or modern processing techniques, thereby the storage life of the produce is enhanced.

Value Added Products

- Fruit Juice: It is a natural juice obtained by pressing out the fruits. Fruit juices may be sweetened or unsweetened.

- RTS: It is prepared from fruit juices which must have atleast 10 per cent fruit juice and 10 per cent total sugar.

- Fruit Juice Powder: The fruit juice is converted into highly hygroscopic powder. These are kept freeze dried and used for fruit juice drinks by reconstituting their composition.

- Fermented fruit beverages: These are prepared by alcoholic fermentation by yeast of fruit juice. The product thus contains varying amounts of alcohols e.g.; Grape wine, orange wine and berry wines from strawberry, blackberry etc.

- Jam: Jam is a concentrated fruit pulp processing a fairly heavy body form rich in natural fruit flavour. It is prepared by boiling the fruit pulp with sufficient quantity of sugar to a reasonably thick consistency to hold tissues of fruit in position.

- Jelly: Jelly is a semi solid product prepared by cooking clear fruit extract and sugar.

- Marmalade: It is usually made from citrus fruits and consists of jelly containing shreds of peels suspended.

- Tomato Ketchup: It is prepared from tomato juice or pulp without seeds or pieces of skin. Ketchup should contain not less than 12 per cent tomato solids and 28 per cent total solids.

- Pickles: Food preserved in common salt or in vinegar is called pickle. Spices and oil may be added to the pickle.

Automation in farm machineries and processing is a very important factor considered to reduce losses, achieve faster and better ways of food processing so as to meet the increasing demands of consumers. Food processing is an important operation that contributes immensely in economic development of the states as it is vital in ensuring food availability and security all over the globe. According to researchers the production process of most agricultural food materials is multi operational process, comprises of many different unit operations requiring separate equipment.

The present trend in agricultural food manufacturing and processing is focused on automation of integrated system that combines many batch operations into single manufacturing system. The design provides on-line and continues control capacity. According to researchers most of the automation systems carried out in food industries are isolated, batch-type operations targeting a specific task. Steve, reported that food processing and manufacturing operations in small and median enterprises (SME) are basically carried out manually unlike in larger industries where automations were achieved either with the aid of robots or combination of simpler electromechanical devices.

Food Processing has been defined as the alteration of raw food materials into consumable state or the later into other forms. According to researchers food processing involves the use of clean raw materials either from crops or animal product to produce good-looking and profitable products and animal feed. It also helps in extending the shelf-life of these products. Food processing removes toxic materials from food, enhanced preservation, marketing, increased food concentration and availability of several foods which are beneficial to the consumers. Health standard of certain group of people with specific health problems such as diabetes and allergies can be improved through modern food processing. Also additional nutrients can be added to certain class of food that lacks such nutrients. It sometimes involves mechanical processes that employ the use of mixing and grinding equipment and machines in the production line. In food processing industry, the food performance parameters are vital element necessary in the design process. Some of these parameters include: hygiene, energy efficiency, labor used.

Importance of Automation in Food Processing

According to studies some of the advantages of automation in food processing include improved productivity in processing line by allowing efficient schedule of work flow and labor utilization. Also it ensures high quality products consistently thereby encourage customer loyalty and this result to expanding market share. In addition one of the major advantages of automation is ensuring food hygiene and safety by eliminating human interference with food product.

For successful automation of agricultural food manufacturing and processing an integration of the manufacturing process must be carried out with view of transforming the operations into single manufacturing design.

Automation and Grain Processing

The grain industry serves as an example of industrial process control, where standards are maintained at a constant rate for product delivery. Automation in grain processing provides both safety and efficiency by supporting the product delivery process. Historically, the grain production process has been hazardous, with industry workers subject to harsh conditions and potential risks associated with the grain, such as flammable grain dust. The transition from manual labor to automation has become the industry's essential shift, with automation now a staple in harvesting, milling, and handling grain. The developers of automation components seek to mitigate production ineffectiveness and inventory miscalculations. Solids level transmitters, responsible for continuously monitoring the amount of grain in potentially hazardous environments, are a reliable replacement for employees at a silo's peak.

Grain processing benefits from automation and industrial wireless.

Thanks to developments in technology, the implementation of sustainable automation does not need to come at the expense of company profit. A sole driver needs to control automation operations when dealing with grain handling, meaning that the number of employees put at risk via that stage in the process is substantially reduced. In order to keep track of inventory, automated management tracks not only the levels of grain inside silos but also where materials are located throughout the production facility. Grain sensors allow for the constant communication of how much grain is being moved through grain elevators and terminals. Instrumentation measures level, weight, and flow of solid grain while maintaining process protection.

Grain terminals allow for grain to be unloaded into hoppers, and then conveyors transfer those hoppers to the elevators. Cleaning drying, and blending machinery all employ automation, reducing the risks presented to employees. Truck and rail load-outs need to be close to target levels, because overloading or under-loading transport results in product loss. Preventing the loss of time and physical resources is a key element of automation. Radar transmitters cut through dust in silos to deliver reliable information back to the process operators. Ultrasonic instrumentation matches with point level technology to indicate when grain levels are high or low.

The handling and blending of raw materials can be monitored by solid flowmeters, with high accuracy still applicable in more compact spaces. A similar solid flowmeter ensures the accuracy of the flow rate and weight of solids measured in bulk, and can also be used in tandem with flow and weigh feeders to keep each individual load of grain consistently accurate. The applications of automation throughout the grain production process have evolved into a reliable means of reducing employee risk while ensuring accuracy and increasing throughput.

Grain operations cover large areas. A reliable, flexible, and cost saving way to establish the needed process control connections throughout the facility is via industrial wireless. Connections between measurement instruments and control units can be created across very long distances without the need to install conduit and cabling. Connections can be created quickly and reliably, with flexibility to increase throughput as more stations or sensors are needed. Equipment is suitable for general or hazardous locations.

Dehulling Systems

Proper treatment and preparation of seeds is a major prerequisite for the effectiveness of subsequent processing. It also influences the quality of final products: oil, press cakes, or extrudate.

Quick and efficient livestock feeding requires the use of protein-rich feed, low in fibre. Hulling of the seeds is a smart solution for this issue.

Hulling technology serves for partial removal of the hull from sunflower seeds and soybeans. Fibre content in the hulls is considerable, especially in the aforementioned types of oilseeds. Removing of at least a fraction of hulls in the processed material leads to a significant decrease of the overall fibre content in press cakes. Another advantage of hulling is higher oil yield during pressing.

Hulling technology is offered at capacities ranging from 600 kg per hour (for the product "Compact") to tens of tons per hour.

The technology also comes with complete engineering and other services, such as technical support, service possibilities, and accessible spare parts.

Sorting Systems

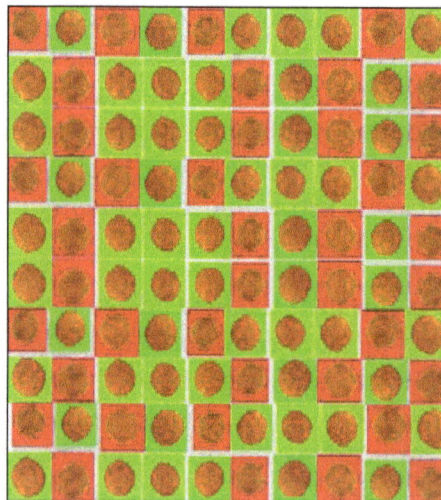

Agricultural products are graded based on their dimensions and quality. This grade is used to sort them and assign them to different sales channels. Each item may yield better income based when properly allocated according to its exact characteristics. Usually, higher grade and bigger agricultural products generate larger revenues. Traditional grading was human-dependent. Later, mechanical devices were used to differentiate agricultural products based on their dimensions and weight. Such devices are still in use today as a reliable method for grading and sorting. More recently, as image processing algorithms emerged, visual inspection techniques were added to perfect the process. They often were manually-tuned and provided a substitute to the human eye, enabling to detect many defects, which humans cannot detect when pace becomes faster. The new wave of intelligent algorithms for grading and sorting is much more powerful than traditional visual analysis algorithms: they have automatic learning capabilities, which ensure a detection performance far beyond the speed and accuracy of any trained operator.

Mechanical Sorters

Mechanical sorters are machines, usually integrated to a conveyor belt, over which agriculture products are sorted by external criteria like dimensions and weight. Such equipment is based on mechanical apparatus triggered by these criteria. For example, a product, be it fruit or vegetable, is dropped into a bucket when its weight or diameter are measured at higher values than a given threshold. When values are lower, it simply travels on the conveyor belt towards the following test. Mechanical sorters are fast and reliable. However, they are limited in that they test only generic criteria.

Challenges

There are many other aspects that should be considered when sorting and grading agricultural product. One example is our modern environment. Today, many people are living in large agglomerations, far from the places where agriculture products are grown, which is sometimes overseas. Such distance calls for longer shelf life, but time is a strong spoilage agent and only fruits and vegetables of the highest quality will survive the longer time to market required. We can predict the potential for preservation in fresh produce by inspecting them for existing defects. Even local defects can expand in time and spoil the whole fruit, making early defect detection a crucial step. Visual defect detection aided by algorithm-based visual systems is used today all over the world. Its goal is to detect as large a set of defect as possible. Such defects can be spotted under the form of color variations, local scratches, bumps and irregular shapes. Many systems today are able to detect those defects. They are usually tuned in advance by an expert operator. The better the skills and patience of the expert, the better will be the quality of detection. This is a real industry challenge, since grading performance may vary depending on personal skills: when these are below par, the results may be profit losses and damage to the reputation of the producer or the distributor.

Optical and Visual Grading

Optical and visual grading systems consist in fruit images being analyzed by dedicated algorithms. Usually, multiple images per a single fruit are used, in order to ensure consistent performance. The analysis algorithms look for defects over the following variables: size, shape, color, and external quality. External quality is detected by color and pattern variations from the standard color and texture. Optical and visual grading can be used to explore the fruit internals by using dedicated cameras, like IR cameras or other cameras responding to a special spectrum.

Existing Systems and Algorithms

Agriculture produce is analyzed today by classical image processing algorithms. Shape, color, and pattern detection algorithms are common. In addition, normalization and equalization algorithms are used to prepare the images for automatic visual inspection. We may find multi-threshold algorithms to isolate phenomena along geometric contour trackers. Such algorithms may detect the defects and clearly bound them to evaluate their severity. All the discussed algorithms require a tuning effort to ensure accurate performance with minimal false or miss-detection. The actual performance is highly dependent on the quality of the visual system and on the effort done in the tuning. A higher skilled operator ensures better algorithm tuning and performance.

New Methods: Deep Learning

Modern methods, such as deep learning, successfully challenge the human factor in traditional vision algorithms. The tuning phase is replaced by automatic learning. When the deep learning algorithm is provided with a set of "good" fruits (oranges for example) and another set of oranges with defects – it is self-adjusted to classify (grade) additional oranges based on the sample sets. No fine tuning is needed here. Every time an orange looks like one of the sets, it is classified and graded accordingly. This method is fast and reliable; more importantly, it yields a consistent performance. The deep learning is the state-of-the-art solution we recommend today for many applications of this kind.

RSIP Vision in Precision Agriculture

RSIP Vision, a powerhouse of vision algorithms, is of course familiar with the most advanced deep learning technology. For several years now, RSIP Vision is using deep learning in detection and classification tasks, like assessing medical pathologies (in eye care and other health care fields) or analyzing written documents in OCR projects.

Computer vision systems are being used increasingly in the field of agricultural and food products for quality assurance purpose. The system offers the generation of precise descriptive data and reduction of tedious human involvement. Computer vision system has proven successful for the objective, online measurement of several agricultural and food products with application ranging from routine inspection to the computer vision system guided robotic control. Some of the areas where the techniques have been applied in agricultural and food processing which need to be developed further for commercial purposes include.

Fruits

External qualities i.e., sizes, shapes and colour are considered of paramount importance in the marketing and sale of fruit. Presence of blemishes influences consumer perceptions and therefore determines the level of acceptability prior to purchase. Computer vision system has been used for the automated inspection and grading of fruit to increase product throughput.

The number of fruits (ripe and unripe) on a tree has been counted by image analysis prior to harvesting. This process involved the development of fruit location algorithms . Images were taken from a distance of about 150cm and all the fruit on the tree placed within the vision field of a camera were counted. The visible fruits were considered to be those that the human eye can distinguish on the monitor. Algorithms based on the red/green relation and on threshold achieved the highest detection percentages on citrus fruit.

The adoption can also be extended to the peasants through tractor hiring bodies or by forming cooperatives to acquire such devices, which only need to be attached to a tractor. The problem of unplanned fruit trees, which may make machine movement very difficult, will also need to be addressed. The high losses incurred presently on fruits in the market comes as a result of the damages and bruises as well as potential damage region inflicted on the fruits falling from the trees due to the method of harvesting of shaking the trees or the branches. The adoption can also be extended to the peasants through tractor hiring bodies or by forming cooperatives to acquire such devices, which only need to be attached to a tractor. The problem of unplanned fruit trees, which may make machine movement very difficult, will also need to be addressed. The high losses incurred presently on fruits in the market comes as a result of the damages and bruises as well as potential damage region inflicted on the fruits falling from the trees due to the method of harvesting of shaking the trees or the branches.

Strawberry appearance and fruit quality are dependent on a number of pre- and post-harvest factors, hence variation occurs, necessitating the need for sorting. Researcher investigated the use of computer vision to sort fresh strawberries, based on size and shape. The experimental results show that the developed system was able to sort the 600 strawberries tested with an accuracy of 94-98% into three grades based on shape and five grades on size.. Average shape and size accuracies of 98 and 100%, respectively, were obtained regardless of the fruit orientation angle with judgement time within 1.18 s.

Prior to export, papayas are subjected to inspection for the purpose of quality control and grading. For size grading, the fruit is weighted manually hence the practice is tedious, time consuming and labour intensive. Therefore, a computer vision system for papaya size grading using shape characteristic analysis. The shape characteristics consisting of area, mean diameter and perimeter were extracted from the papaya images. Some researchers classified according to combinations of the three features to study the uniqueness of the extracted features. The proposed technique showed the ability to perform papaya size classification with more than 94% accuracy.

Vegetables

The need to be responsive to market demands places a greater emphasis on quality assessment resulting in the greater need for improved and more accurate grading and sorting practices. Computer vision system has shown to be a viable means of meeting these increased requirements for the vegetables field.

Plantlet segments of potato were subcultures for classification by colour machine vision system by Alchanatis. In related development, two algorithms were developed for analyzing digital binary images and estimating the location of stem root joints in processing carrots. Both algorithms were capable of estimating the stem/root location with a standard deviation of 5mm. Also Haworth and Searcy classified carrots on surface defects, curvature and brokenness. A line scans images with discrete Fourier transform was developed for the classification of broccoli heads for assessing its maturity. For the 160 observations from each of three broccolis cultivates, an accuracy of 85% was achieved for multiple cultivars.

Mushrooms' discolouration is undesirable in mushroom houses and it reduces market value. The colour and shape of the cap is the most important consideration of fresh mushrooms. Felfodi and Vizhanyo used mushroom images recorded by a machine vision system to recognize and identify discolouration caused by bacterial disease. The method identified all the diseased spot as 'diseased' and none of the healthy mushrooms parts were detected as 'diseased'. Reed used camera-based technology to select mushroom by size for picking by a mushroom harvester.

Potatoes have many possible shapes, which need to be graded for sale into uniform classes for different markets. This created difficulties for shape separation. A Fourier analysis based shape separation method for grading of potatoes using machine vision for automated inspection was developed by A shape separator based on harmonics of the transform was defined. Its accuracy of separation was 89% for 120 potato samples, in agreement with manual grading. Earlier, studied the use of computer vision for locating the position of pulp extraction automatically for the purpose of further analysis on the extracted sample. An image acquisition system was also constructed for mounting on a sweet potato harvester for the purpose of yield and grade monitoring. It was found that culls were differentiated from saleable sweet potatoes with classification rates as high as 84%.

Chilli is a variety grown extensively consumed by almost all the population. It has a high processing demand and proper sorting is required before filling or canning. A sorter that, Federico Hahn classifies chilli by three different width sizes was built. The conveyor used baby suckers to align each chilli during sensing. Chilli width was determined by means of a photodiode scanner, which detected the incoming radiation sent by a laser line generator. Chilies presenting necrosis were detected with a radiometer and removed to increase product quality. Horizontal and vertical widths were measured for 200 chilies. The accuracy on the necrosis detection and width classification was of 96.3 and 87%, respectively. On-line necrosis measurements were 85% accurate when only the relative reflectance at 550nm was used.

Vegetable quality is frequently referred to size, shape, mass, firmness, colour and bruises from which fruits can be classified and sorted. However, technological by small and middle producers implementation to assess this quality is unfeasible, due to high costs of software, equipment as well as operational costs. Based on these considerations, the Antonio Carlos Loureiro research is to evaluate a new open software that enables the classification system by recognizing fruit shape, volume, colour and possibly bruises at a unique glance. The software named ImageJ, compatible with Windows, Linux and MAC/OS, is quite popular in medical research and practices, and offers algorithms to obtain the above-mentioned parameters. The software allows calculation of volume, area, averages, border detection, image improvement and morphological operations in a variety of image archive formats.

Some other earlier studies of computer vision associated with vegetable grading and inspection include colour and defect sorting of bell peppers. presented the techniques of vision inspection of mushrooms, apples and potatoes for size, shape and colour. The use of computer vision for the location of stem/root joint in carrot has also been assessed. Feature extraction and pattern recognition techniques were developed by Howarth and Searcy to characterize and classify carrots for forking, surface defects, curvature and brokenness. The rate of misclassification was reported to be below 15% for the 250 samples examined. More recently sweet onions were line scanned for internal defects using X-ray imaging . An overall accuracy of 90% was achieved when spatial and transform features were evaluated for product classification.

Image analysis and machine vision in general from the foregoing can be said to offer a fast and reliable process in product sorting and grading, separation and detection of some other facilities.

Grains

Grain quality attributes are very important for all users and especially the milling and baking industries. Computer vision has been used in grain quality inspection for many years. An early study by used machine vision to identify different varieties of wheat and to discriminate wheat from non-wheat components.

In later research found that wheat classification methods could be improved by combining morphometry (computer vision analysis) and hardness analysis. Hard and soft recognition rates of 94% were achieved for the seventeen varieties examined. Twenty-three morphological features were used for the discriminant analysis of different cereal grains using machine vision . Classification accuracies of 98, 91, 97,100 and 91% were recorded for CWRS (Canada Western Red Spring) wheat, CWAD (Canada Western Amber Durum) wheat, barley, oats and rye, respectively. 25 kernels per image were captured from a total of 6000 for each grain type examined.The relationship between colour and texture features of wheat samples to scab infection rate was studied using a neural network method . It was found that the infection rates estimated by the system followed the actual ones with a correlation coefficient of 0.97 with human panel assessment and maximum and mean absolute errors of 5 and 2%, respectively. In this study machine vision-neural network based technique proved superior to the human panel. Image analysis has also been used to classify dockage components for CWRS (Canada Western Red Spring) wheat and other cereals . Morphology, colour and morphology-colour models were evaluated for classifying the dockage components. Mean accuracies of 89 and 96% for the morphology model and 71 and 75% for the colour model were achieved when tested on the test and training data sets, respectively. Overall 6000 kernels for each grain type were analyzed. Machine vision was used to identify weeds commonly found in wheat fields in experimentation by Zhang and Chaisattapagon . Five shape parameters were used in leaf shape studies and were found effective in distinguishing broadleaf weed species such as pigweed, thistle and kochia from wheat.

In order to preserve corn quality it is important to obtain physical properties and assess mechanical damage so as to design optimum handling and storage equipment. Measurements of kernel length, width and projected area independent of kernel orientation have been performed using machine vision . The algorithm accuracy was between 0.86 and 0.89 measured by the correlation coefficient between predicted results and actual sieving for a 500 g sample. The processing time of the size- grading program was reported as being between 0.66 and 0.74 s per kernel. Steenhoek and Precetti performed a study to evaluate the concept of two-dimensional image analysis for classification of maize kernels according to size category. A total of 320 maize kernels were categorized into one of 16 size categories based on degree of roundness and flatness. Classification accuracy of both machine vision and screen systems was above 96% for round-hole analysis. However, sizing accuracy for flatness was less than 80%.

Ng et al. developed a machine vision algorithm for corn kernel mechanical and mould damage measurement, which demonstrated a standard deviation less than 5% of the mean value of the 250 grains examined. They found that this method was more consistent than other methods available. The automatic inspection of 600 corn kernels was also performed by using machine vision. For whole and broken kernel identification on-line tests had successful classification rates of 91 and 94% for whole and broken kernels, respectively.

The whiteness of corn has been measured by an on-line computer vision approach by

Liu and Paulsen. For the 63 samples (50- 80 kernels per sample) tested the technique was found to be easy to perform with a speed of 3 kernels per s. In other studies Xie and Paulsen used machine vision to detect and quantify tetrazolium staining in corn kernels. The tetrazolium-machine vision algorithm was used to predict heat damage in corn due to drying air temperature and initial moisture content.

As rice is one of the leading food crops of the world its quality evaluation is of importance to ensure it remains appealing to consumers. Liu et al. developed a digital image analysis method for measuring the degree of milling of rice. They compared the method with conventional chemical analysis and obtained a coefficient of determination of R2 =/0.9819 for the 680 samples tested. employed three online classification methods for rice quality inspection:-namely range selection, neural network and hybrid algorithms. The highest recorded online classification accuracy was around 91% at a rate of over 1200 kernels/min. The range selection method achieved this accuracy but required time-consuming and complicated adjustment. In another study, milled rice from a laboratory mill and a commercial-scale mill was evaluated for head rice yield and percentage whole kernels, using a shaker table and a machine- vision system called the Grain Check.

3-D Technique

In general, only 2-dimensional (2D) data are needed for grading, classification, and analysis of most agricultural images. However, in many applications 3-dimensional image analysis maybe needed as information on structure or added detail is required. A 3-D vision technique has been developed to derive a geometric description from a series of 2-D images . In practice this technique might be useful for food inspection. For example, when studying the shape features of a piece of bakery, it is necessary to take 2-D images vertically and horizontally to obtain its roundness and thickness, respectively.

Researchers investigated the feasibility of using a charge simulation method (CSM) algorithm to process primary image features for three dimensional shape recognition. The required features were transferred to a retinamodel identical to the prototype artificial retina and were compressed using the CSM by computing output signals at work cells located in the retina. An overall classification rate of 94% was obtained when the prototype artificial retina discriminated between distinct shapes of oranges for the 100 data sets tested. Gunasekaran and Ding obtained 3- D images of fat globules in cheddar cheese from 2-D images. This enabled the in situ 3-D evaluation of fat globule characteristics so as the process parameters and fat levels may be changed to achieve the required textural qualities.

Problems of Automated Sorting and Grading

The major problem of automated sorting and grading is one of socio-economic effects, which tend to reduce employment when the number of operators required in the processing line is reduced. It is not suitable in processes where manual skill is necessary or economically more attractive. It requires higher initial and maintenance cost and there may be the need for a precise understanding of the process for programming to achieve the required product quality Equipment associated problem involves radiance on accurate sensors to precisely measure process condition and the increased risk, delays and cost if the automatic system fails. The farm layout and very low production level which may make it uneconomically viable has also been highlighted.

Prospects of Automated Sorting and Grading

The adoption of this emerging technology by first putting more effort into researches on the appropriate methods and ways of application will be of immense benefit to this country. Some of the other associated benefits include increased production rates (e.g. through optimization of equipment utilization), more efficient operation, production of more consistent product quality, greater product stability and safety.

With the above in mind, the fruit and vegetable processing and marketing industries stand to gain from this emerging technology. This is because the losses incurred during the harvesting season on fruit are enormous. This results from large number of products to be handled and sold at the same time. These products consist of ripe and unripe fruits. The introduction of automated sorting and grading will encourage the sorting of the unripe (which can be kept for a relatively longer period) from the ripe, which are to be sold immediately. Presently the practice at the fruit market in to sell baskets of fruits containing both ripe and unripe. These products are found to get spoilt before they get to the final destination.

One of the major hindrances to the introduction of this technique is the level of production, which remains at the peasant level. Most of the products found in the markets are owned by a number of individuals with each controlling not more than 4 to 5 baskets, which will be too small for the adoption of an automated technique. However, encouraging wholesalers and retailers for groups and cooperatives in marketing as practiced in the village level for maize shelling will be a way out.

The adaptation of computer vision system for quality evaluation of processed foods is the area for the greatest potential uptake of this technology, as analysis can be based on a standard requirement in already automated controlled conditions. More complex systems are needed for the automated grading and sorting of fresh produce because of the greater range in variability of quality and also as produce orientation may influence results. With the idea of precision and more environmental friendly agriculture becoming more realistic the potential for computer vision system in this area is immense with the need in field crop monitoring, assessment and guidance systems. However, techniques such as 3D and colour vision will ensure computer vision development continues to meet the accuracy and quality requirements needed in this highly competitive and changing industry.

References

- Uses-of-conveyor-belting-in-the-agricultural-industry: mayconveyor.com, Retrieved 19 June, 2019

- Grain-drying: ag.ndsu.edu, Retrieved 15 March, 2019

- Automation-of-integrated-system-for-grain-beverages-processing, automation-in-agriculture-securing-food-supplies-for-future-generations: intechopen.com, Retrieved 07 January, 2019

- Automation-and-grain-processing: blog.analynk.com, Retrieved 27 February, 2019

- Dehulling-and-separation-of-hulls: farmet.cz, Retrieved 02 January, 2019

- Grading-and-sorting: rsipvision.com, Retrieved 23 March, 2019

Agricultural Robots

Agricultural robots are used to increase the production yield for farmers. Driverless tractor, weeding robots, fruit picking robots, harvest automation robot, agricultural drones, etc. are some of these robots. This chapter discusses about these agricultural robots in detail.

Agriculture is quickly becoming an exciting high-tech industry, drawing new professionals, new companies and new investors. The technology is developing rapidly, not only advancing the production capabilities of farmers but also advancing robotics and automation technology An agbot, also called an agribot, is an autonomous robot used in farming to help improve efficiency and reduce reliance on manual labor. Future farms are expected to be tilled, sown, tended and harvested solely by fleets of co-operating autonomous robots called swarm robots that will weed, fertilize, control pests and diseases, all the while collecting valuable data.

Equipped with specialized arms, end effectors and other tools to perform a variety of agricultural tasks, agbots can connect to wireless sensor networks (WSNs) and with the help of drones, gather large amounts of data. Big data analytics will help farmers extract information from the vast amount of data to make farming more efficient and improve output.

The trend toward using agbots is expected to increase exponentially to help alleviate a shortage of human farm labor. Today, robots for fruit picking, milking and sheep shearing have successfully done jobs that previously required human manual labor. Thousands of robotic milking parlors are in place worldwide, and mobile bots are helping dairy farmers automate tasks such as feed pushing and cleaning manure.

Using Agbots:

- Equipment manufacturers are developing tractors that use machine vision and autonomous robotics to replace traditional large, expensive and inefficient tractors.

- In the rules-oriented organic farming field, robotic weeding implements have been used to inspect crop rows and identify weeds rapidly.

- Robots operating at low speeds can detect weeds. Next-generation computer vision algorithms will be used to classify them.

- Robots are being used for harvesting and sorting, packaging and boxing.

- Other future applications include planting seeds, grafting plants, seeding clouds, analyzing soil and monitoring the environment.

Fruit picking robots, driverless tractor/sprayers, and sheep shearing robots are designed to replace human labor. In most cases, a lot of factors have to be considered (e.g., the size and color of the fruit to be picked) before the commencement of a task. Robots can be used for other horticultural tasks such as pruning, weeding, spraying and monitoring. Robots can also be used in livestock applications (livestock robotics) such as automatic milking, washing and castrating. Robots like these have many benefits for the agricultural industry, including a higher quality of fresh produce, lower production costs, and a decreased need for manual labor. They can also be used to automate manual tasks, such as weed or bracken spraying, where the use of tractors and other manned vehicles is too dangerous for the operators.

Designs

Fieldwork robot.

The mechanical design consists of an end effector, manipulator, and gripper. Several factors must be considered in the design of the manipulator, including the task, economic efficiency, and required motions. The end effector influences the market value of the fruit and the gripper's design is based on the crop that is being harvested.

End Effectors

An end effector in an agricultural robot is the device found at the end of the robotic arm, used for various agricultural operations. Several different kinds of end effectors have been developed. In an agricultural operation involving grapes in Japan, end effectors are used for harvesting, berry-thinning, spraying, and bagging. Each was designed according to the nature of the task and the shape and size of the target fruit. For instance, the end effectors used for harvesting were designed to grasp, cut, and push the bunches of grapes.

Berry thinning is another operation performed on the grapes, and is used to enhance the market value of the grapes, increase the grapes' size, and facilitate the bunching process. For berry thinning, an end effector consists of an upper, middle, and lower part. The upper part has two plates and a rubber that can open and close. The two plates compress the grapes to cut off the rachis branches and extract the bunch of grapes. The middle part contains a plate of needles, a compression spring, and another plate which has holes spread across its surface. When the two plates compress, the needles punch holes through the grapes. Next, the lower part has a cutting device which can cut the bunch to standardize its length.

For spraying, the end effector consists of a spray nozzle that is attached to a manipulator. In practice, producers want to ensure that the chemical liquid is evenly distributed across the bunch. Thus, the design allows for an even distribution of the chemical by making the nozzle to move at a constant speed while keeping distance from the target.

The final step in grape production is the bagging process. The bagging end effector is designed with a bag feeder and two mechanical fingers. In the bagging process, the bag feeder is composed of slits which continuously supply bags to the fingers in an up and down motion. While the bag is being fed to the fingers, two leaf springs that are located on the upper end of the bag hold the bag open. The bags are produced to contain the grapes in bunches. Once the bagging process is complete, the fingers open and release the bag. This shuts the leaf springs, which seals the bag and prevents it from opening again.

Gripper

The gripper is a grasping device that is used for harvesting the target crop. Design of the gripper is based on simplicity, low cost, and effectiveness. Thus, the design usually consists of two mechanical fingers that are able to move in synchrony when performing their task. Specifics of the design depend on the task that is being performed. For example, in a procedure that required plants to be cut for harvesting, the gripper was equipped with a sharp blade.

Manipulator

The manipulator allows the gripper and end effector to navigate through their environment. The manipulator consists of four-bar parallel links that maintain the gripper's position and height. The manipulator also can utilize one, two, or three pneumatic actuators. Pneumatic actuators are motors which produce linear and rotary motion by converting compressed air into energy. The pneumatic actuator is the most effective actuator for agricultural robots because of its high power-weight ratio. The most cost efficient design for the manipulator is the single actuator configuration, yet this is the least flexible option.

Development

The first development of robotics in agriculture can be dated as early as the 1920s, with research to incorporate automatic vehicle guidance into agriculture beginning to take shape. This research led to the advancements between the 1950s and 60s of autonomous agricultural vehicles. The concept was not perfect however, with the vehicles still needing a cable system to guide their path. Robots in agriculture continued to develop as technologies in other sectors began to develop as well. It was not until the 1980s, following the development of the computer, that machine vision guidance became possible.

Other developments over the years included the harvesting of oranges using a robot both in France and the US.

While robots have been incorporated in indoor industrial settings for decades, outdoor robots for the use of agriculture are considered more complex and difficult to develop. This is due to concerns over safety, but also over the complexity of picking crops subject to different environmental factors and unpredictability.

Demand in the Market

There are concerns over the amount of labor the agricultural sector needs. With an aging population, Japan is unable to meet the demands of the agricultural labor market. Similarly, the United State currently depends on a large number of immigrant workers, but between the decrease in seasonal farmworkers and increased efforts to stop immigration by the government, they too are unable to meet the demand. Businesses are often forced to let crops rot due to an inability to pick them all by the end of the season. Additionally, there are concerns over the growing population that will need to be fed over the next years. Because of this, there is a large desire to improve agricultural machinery to make it more cost efficient and viable for continued use.

Current Applications and Trends

Much of the current research continues to work towards autonomous agricultural vehicles. This research is based on the advancements made in driver-assist systems and self-driving cars.

While robots have already been incorporated in many areas of agricultural farm work, they are still largely missing in the harvest of various crops. This has started to change as companies begin to develop robots that complete more specific tasks on the farm. The biggest concern over robots harvesting crops comes from harvesting soft crops such as strawberries which can easily be damaged or missed entirely. Despite these concerns, progress in this area is being made. According to Gary Wishnatzki, the co-founder of Harvest Croo Robotics, one of their strawberry pickers currently being tested in Florida can "pick a 25-acre field in just three days and replace a crew of about 30 farm workers". Similar progress is being made in harvesting apples, grapes, and other crops.

Another goal being set by agricultural companies involves the collection of data. There are rising concerns over the growing population and the decreasing labor available to feed them. Data collection is being developed as a way to increase productivity on farms. AgriData is currently developing new technology to do just this and help farmers better determine the best time to harvest their crops by scanning fruit trees.

Types of Field Robots used in Agriculture Applications

There are many different types of field robots used in agriculture and new solutions are in development every day. The following 8 types of field robots have become the most popular:

- Precision agriculture: These field robots are used on small farms or vineyards and enable precision agriculture techniques. Often, they're used to autonomously monitor soil respiration, photosynthetic activity, leaf area indexes (LAI) and other biological factors.

- Pollution monitoring: Some field robots are now equipped to monitor the pollution created by agriculture at the ground level. These robots measure carbon dioxide and nitrous oxide emissions so that farmers can reduce their environmental footprint.

- Livestock ranching: A new type of field robot is used to herd livestock on large ranches. These robots also monitor the animals and ensure they're healthy and have enough area to graze.

- Weed control: Field robots for weed control can autonomously navigate a farm and deliver targeted sprays of herbicides to eliminate weeds. This approach reduces crops' exposure to herbicides and helps prevent the growth of herbicide-resistant weeds.

- Nursery automation: Field robots can be used in crop nurseries, primarily to move plants around large greenhouses. These robots create major efficiencies for crop nurseries and help address a growing labor shortage.

- Crop harvesting: For harvesting crops, specialized field robots can work around the clock for faster harvesting, in some cases completing the same amount of work as approximately 30 workers.

- Fruit harvesting: Field robots are starting to be used to harvest fruit in addition to crops. Fruit harvesting is notoriously difficult for robots. These field robots are equipped with advanced vision systems to identify fruits and grasp them without damaging them.

- Planting and seeding: An emerging application, field robots with 3D vision systems can now accurately plant and seed crops for optimal growth, primarily for lettuce farming and vineyards.

Field robots are diverse in their form and function, even within the agriculture industry. As farmers face a number of growing challenges, field robots are stepping up and helping them improve efficiency and bring in larger crop yields.

The robots can be used on farms and they enable precision agriculture techniques, they are used to autonomously monitor soil respiration, photosynthetic activity, leaf area indexes (LAI) and other biological factors, The robots are used to herd livestock on large ranches, They monitor the animals and ensure they are healthy and have enough area to graze.

The robots are equipped to monitor the pollution created by agriculture at the ground level, They can measure carbon dioxide & nitrous oxide emissions, so, the farmers can reduce their environmental footprint, Crop Harvesting robots can work around the clock for faster harvesting, in some cases completing the same amount of work as approximately 30 workers.

Robots offer an efficient method in spraying pesticides, Micro-spraying robots can autonomously navigate the farm and deliver targeted sprays of herbicides to eliminate weeds, This approach reduces crops' exposure to herbicides and helps prevent the growth of herbicide-resistant weeds.

Fruit Harvesting robots are used to harvest fruit in addition to crops, Fruit harvesting is notoriously difficult for robots, These robots are equipped with the advanced vision systems to identify fruits and they can grasp them without damaging them, The robots with 3D vision systems can accurately plant & seed crops for optimal growth.

Nursery Automation robots can create major efficiencies for crop nurseries, primarily to move plants around large greenhouses. They help address a growing labor shortage. Nursery automation is where seeds are grown into young plants, which are later planted outside. Nursery plants are sold directly to consumers and landscape gardeners.

The robots can be attachments to a tractor. As humans drive the tractors, the robots are designed to adapt to the speed that the human is driving. However, fully-autonomous tractors are also becoming popular. Most agricultural robots are applied in crop growing. There have been a few emerging applications within sheep & cattle farming. Collaborative robots are used to help in the milking process on dairy farms. The UR5 can be used to spray disinfectant on the cow's udders in preparation for milking.

Robots are used in crop seeding. Autonomous precision seeding combines robotics with geomapping. The map is generated which shows the soil properties (quality, density, etc) at every point in the field. The tractor, with robotic seeding attachment, places the seeds at precise locations and depths so that each has the best chance of growing.

Robots are used in crop monitoring and analysis. New sensor and geomapping technologies enable the farmers to get a much higher level of data about their crops than they have in the past. Ground robots and drones provide a way to collect this data autonomously.

The farmer can move the drone to the field, and initiate the software via a tablet or smartphone, and view the collected crop data in real time. Ground-based robots provide more detailed monitoring as they can get closer to the crops. Some can be used for other tasks like weeding and fertilizing.

Robot-assisted Precision Irrigation can reduce wasted water by targeting specific plants. Ground robots autonomously navigate between rows of crop and pour water directly at the base of each plant. Robots are used in picking & harvesting crops. It can be done with a combine harvester, which can be automated just like a tractor.

Robots can access areas where other machines cannot. Robots present a higher quality of fresh products, lower production costs, and a decreased need for manual labor. They can be used to automate manual tasks. Thinning robot uses computer vision to detect plants as it drives over them and decides which plants to keep and which to remove. Thinning reduces the density of plants so that each has a better chance of growing. Pruning involves cutting back parts of plants to improve their growth.

Weeding robots are used in weeding the plants. They use computer vision to detect the plants as it is pushed by the tractor. It automatically hoes the spaces between the plants to uproot the weeds. Other weeding robots use lasers to kill the weeds and they do not need to use chemicals.

Farmers can use fruit picking robots, driverless tractor/sprayers, and sheep shearing robots. They are designed to replace human labor. Robots can be used in pruning, weeding, spraying and monitoring. They can be used in livestock applications (livestock robotics) such as automatic milking, washing and castrating.

Outdoor Agriculture Robotics

Driverless Tractor

A driverless tractor is an autonomous farm vehicle that delivers a high tractive effort (or torque) at slow speeds for the purposes of tillage and other agricultural tasks. It is considered driverless because it operates without the presence of a human inside the tractor itself. Like other unmanned ground vehicles, they are programmed to independently observe their position, decide speed, and avoid obstacles such as people, animals, or objects in the field while performing their task. The various driverless tractors are split into full autonomous technology and supervised autonomy. The idea of the driverless tractor appears as early as 1940, but has really come to fruition in the last few years. The tractors use GPS and other wireless technologies to farm land without the need of a driver. They operate simply with the aid of a supervisor monitoring the progress at a control station or with a manned tractor in lead.

Case IH utilize "follow me" technology and vehicle-to-vehicle communication with a driverless tractor that follows one operated by a person.

The idea of a driverless tractor has been around since as early as 1940, when Frank W. Andrew invented his own. To guide his driverless tractor, a barrel or fixed wheel would be put in the center of the field and around it would wind a cable attached to a steering arm on the front of the tractor. In the 1950s Ford developed a driverless tractor that they called "The Sniffer" but it was never produced because it could not be operated without running wire underground through the field. There were no major advances in driverless tractor technologies until 1994 when engineers at the Silsoe Research Institute developed the picture analysis system, which was used to guide a small driverless tractor designed for vegetable and root crops. This new tractor could even handle slight headland turns.

Current driverless tractor technologies build on recent developments in unmanned vehicles and agricultural technology. A tractor is defined as a powerful motor-driven vehicle with large, heavy treads, used for pulling farm machinery and other vehicles. Most commonly, the term is used to describe a farm vehicle that provides the power and traction to mechanize agricultural tasks. Precision agriculture was a major shift in technology that occurred in the 1980s. The result was tractors that farmers drove with the aid of GPS devices and on-board computers. Precision agriculture focuses on maximizing returns while using minimum resources. With the aid of GPS devices and computers, farmers could use tractors more efficiently.

Next, engineers worked on semi-automated tractors. These tractors had drivers, but the drivers only had to steer at the end of each row. Subsequently, the idea of a driverless tractor emerged in 2011 and 2012.

Driverless tractors were initially created to follow a main tractor (with a driver). This would allow one driver to do twice as much work using what is called "follow-me" technology. The driverless tractor would follow a lead tractor between fields just like a hired hand would. But now driverless tractor technologies have moved toward autonomy, or independent functioning.

Technology

The driverless tractor is part of a move to increase automation in farming. Other such autonomous technologies currently utilized in farming include automatic milking and automatic strawberry pickers. Developing such a technology is difficult. In order for it to be successful, the tractor must be able to follow deterministic tasks (a task that is defined before it starts, such as a path to follow on a field), have reactive behavior (the ability to react to an unknown situation such as an obstacle in the way), and have reflexive responses (making a decision without hesitation or time consuming calculations such as changing the steering angle if necessary). Ultimately, the tractor should imitate a human in its ability to observe spatial position and make decisions such as speed.

Working of Technology

The technology for the driverless tractor has been evolving since its beginnings in the 1940s. There are now several different approaches to building and programming the tractors.

Full Autonomy

Driverless compact tractors perform fully autonomous spraying tasks at a Texas vineyard.

Currently, the majority of fully autonomous tractors navigate using lasers that bounce signals off several mobile transponders located around the field. These lasers are accompanied with 150 MHz radios to deal with line-of-sight issues. Instead of drivers, the tractors have controllers. Controllers are people that supervise the tractors without being inside them. These controllers can supervise multiple tractors on multiple fields from one location.

Another fully autonomous tractor technology involves using the native electrical (or CAN bus) system of the tractor or farm equipment to send commands. Using GPS positioning and radio feedback, automation software manages the vehicle's path and controls farming implements. A retrofit radio receiver and on-board computer are generally used to receive commands from the remote command station and translate it into vehicle commands such as steering, acceleration, braking, transmission, and implement control. Sensor technologies such as lidar improve safety by detecting and reacting to unforeseen obstacles.

Supervised Autonomy

Tractors that function with supervised autonomy (automated technology, but with a supervising operator present) use vehicle-to-vehicle (V2V) technology and communication. There is a wireless connection between the two tractors to exchange and share data. The leading tractor (with an operator) determines speed and direction which is then transmitted to the driverless tractor to imitate.

Safety

The driverless tractor is considered controversial in terms of safety and public acceptance. A tractor operating without a driver makes some people nervous. Creating technology that stays safe in all scenarios where failure could possibly occur takes a lot of programming and time. In terms of motion detection, the tractors have sensors to stop them if they detect objects in their path such as people, animals, vehicles or other large objects.

Manufacturers of Driverless Tractors

There are several primary manufacturers that have been actively seeking to produce a marketable driverless tractor and have made strides toward substantial prototypes and mass production of their products. Current leading manufacturers are John Deere, Autonomous Tractor Corporation, Fendt and Case IH.

John Deere

John Deere has had a strong influence on the development of automated farming technology. In early 2008, Deere and Company launched its ITEC Pro guidance product, an automated system based on global positioning technology which automates vehicle functions including end turns. John Deere currently has a prototype in development. Rather than the use of lasers, the tractor uses two 6-inch dome antennas that receive signals from a global positioning satellite. Based on these satellite signals, the tractor follows a previously programmed route via an electronic map. These antenna are also for human operators to control the tractor if satellite signals have trouble penetrating buildings or heavy vegetation.

Autonomous Tractor Corporation

In January 2012, Terry Anderson established a company called Autonomous Tractor Corporation (ATC) located in North Dakota. The SPIRIT driverless tractor is a product of ATC. Anderson thought tractors were getting bigger and more expensive while not improving in quality. He thus came out of retirement to create the SPIRIT. The tractors Anderson helped create were initially made to follow a main tractor (that has a driver) but are now moving towards independent functioning. Anderson tested half-scale models of his automated tractor design at his second home in Texas. He presented his model at the Big Iron Farm Show in September 2012. Anderson stated that his goal is to build 25 units of his automated tractor in 2013 and sell them at a discount price for farmers to test.

Fendt

Fendt, part of the AGCO corporation, has also been working on a driverless tractor. In 2011 in Hanover at the Agritechnica show, Fendt presented their driverless tractor model called the GuideConnect. The GuideConnect is a tractor programmed to mirror the movements of another tractor. An operator-driven tractor maneuvers through a field or through crops and is followed by a driverless tractor. Because the first tractor has a driver, he or she can manually steer to avoid obstacles with the driverless tractor following. Instead of focusing on a completely independent piece of technology, Fendt made GuideConnect to work together with operator-controlled machinery. GuideConnect is connected by satellite navigation and radio to the operator-driven tractor. The two perform as one unit. The one with the driver leads while the one without the driver follows.

Case IH

Case IH is a company created by the merging of J.I. Case Company and International Harvester. The company now operates under CNH Global, but the tractors are still branded Case IH. The driverless tractors produced by Case IH are referred to as "supervised autonomy." A tractor driven by a person is followed by autonomous machinery which copies the steering and speed of the former tractor. There is an initial driver, but the autonomous technology is present in the second tractor. The two machines operate with V2V technology, which is vehicle-to-vehicle communication. In 2016 Case unveiled their latest autonomous concept, a cabless row crop tractor of the Magnum model that could operate autonomously.

Tractor Upgrade Kits

Autonomous steering systems can also be added to non-driverless tractors to make them driverless.

Spraying and Weeding Robots

Weeds in agricultural production are mainly controlled by herbicides. As in organic farming no herbicides can be used, weed control is a major problem. While there is sufficient equipment available to control the weeds in between the rows, weed control in the rows (intra-row weeding) still requires a lot of manual labour. This is especially the case for crops that are slowly growing and shallowly sown like sugar beet, carrots and onions. In 1998, on average 73 hours per hectare sugar beet were spent on hand weeding in the Netherlands. The required labour for hand weeding is expensive and often not available. An autonomous weeding robot replacing this labour, could mean an enormous stimulus for organic farming.

The Design Procedure

Method

The autonomous weeding robot is designed using a phase model as the design method. In this phase model the design of a product is represented as a process consisting of a problem definition phase, alternatives definition phase and a forming phase. The results of the different phases are solutions on different levels of abstraction.

The problem definition phase starts with defining the objective of the design. In the problem definition phase also the set of requirements is established. The requirements can be split into fixed and variable requirements. A design that does not satisfy the fixed requirements is rejected. Variable requirements have to be fulfilled to a certain extent. To what extent these requirements are fulfilled, determines the quality of the design. The variable requirements are also used as criteria for the evaluation of possible concept solutions. The last part of the problem definition phase consists of the definition of the functions of the robot. A function is an action that has to be performed by the robot to reach a specific goal. In our case, important functions are 'intra-row weeding' and 'navigate along the row'. The functions are grouped in a function structure, which represents a solution on the first level of abstraction.

The function structure consists of several functions. Every function can be accomplished by several alternative principles, e.g. mechanical and thermal principles for weed removal. In the alternatives definition phase, possible alternative principles for the various functions are presented in a morphological chart. The left column lists the functions and the rows display the alternative principles. By selecting one alternative for each function and by combining these alternatives, concept solutions can be established. These concept solutions are represented by lines drawn in the morphological chart. The best concept solution can be selected using a rating procedure. In the forming phase this selected concept solution is worked out into a prototype.

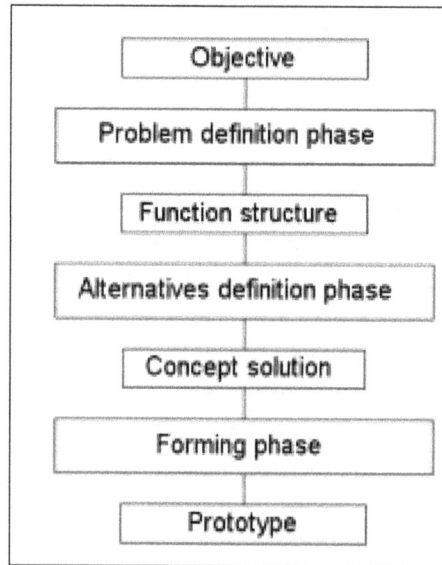

The design process.

Application for the Weeding Robot

The objective of the research is formulated as 'replacement of hand weeding in organic farming by a device working autonomously at field level'. Starting from this objective, the first step in the problem definition phase was to establish the set of requirements. For this purpose interviews were held with potential users, scientists and consultants related to organic farming. The resulting requirements are as follows:

Fixed requirements:

- Replacing hand weeding in organic farming.

- Applicable in combination with other weed control measures.

- Manual control of the vehicle must be possible for moving the vehicle over short distances.

- Weeding a field autonomously.

- Ability to work both day and night.

- The weeding robot should not cross the field boundaries.

- The weeding robot must be self restarting in absence of emergency.

- The weeding robot informs the farmer when the weeding robot stopped definitely (e.g. due to security reasons) or when it is ready.

- The weeding robot sends its operational status to the user at request.

- The weeding robot must function properly in sugar beet.

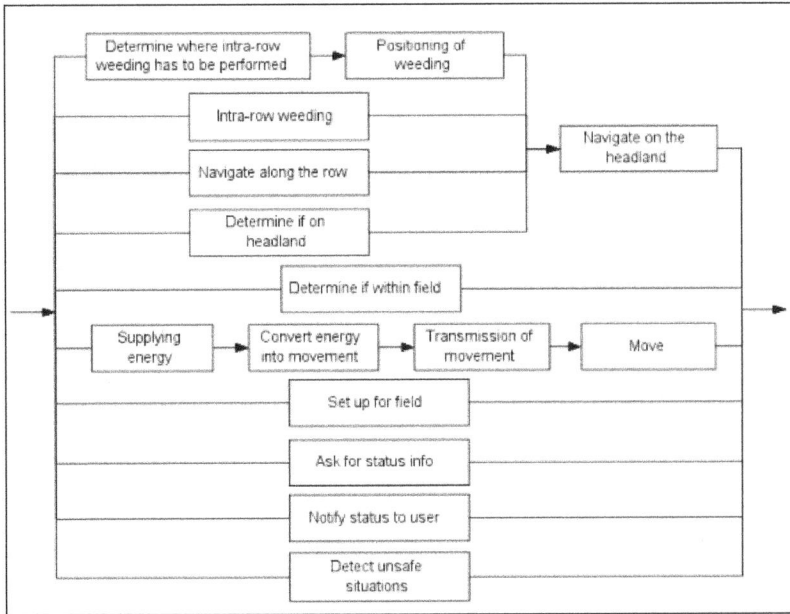

The function structure.

Variable requirements:

- Removing more than 90 percent of the weeds in the row.

- The costs per hectare may be at least comparable to the costs of hand weeding.

- Damage to the crop is as low as possible.

- The soil pressure under the weeding robot must be comparable or less than for hand weeding.

- Energy efficient.

- Safe for people, animals and property.

- Suitable as research platform.

- Limited noise production.

- Reliable functioning.

- Easy to use.

After establishing the set of requirements the functions of the the weeding robot were identified. These functions were grouped into a function block scheme. This scheme is represented in figure. The lines in the scheme indicate flows of energy, material or information. Functions located in parallel lines can be performed simultaneously.

The navigation system consists of four functions. Firstly, the weeding robot should constantly determine if it is located in- or outside the field. Secondly, if within the field, it should determine if it is on one of the headlands or not. Thirdly, in case it is not on the headlands, it should navigate along the row and perform the intra-row weeding. Fourthly, if the weeding robot arrives on the headland, it should stop the intra-row weeding and start to navigate to the next crop rows to be weeded. This sequence repeats until the whole field, except the headlands, is weeded. Weeding of the headlands is left out of consideration. An increasing number of farmers in the Netherlands do not grow sugar beet at the headlands because they think it is not cost-effective.

In the alternatives definition phase possible alternative principles for the various functions are listed in a morphological chart. Four people involved in the project drew lines indicating possible concept solutions in the chart. These concept solutions were then weighed against each other using the variable requirements listed before. The concept solution indicated by the line in figure is the final concept solution.

Results of the Design Process

Determine where Intra-row Weeding has to be Performed

To determine where intra-row weeding has to be performed, pattern recognition of plant locations is going to be used. From earlier research it is expected that the quality of detection of this method is at least as good as the quality of detection of other methods. Though combinations of methods like recognition of pattern, shape and colour are expected to have a potential for higher quality of detection, just pattern recognition is chosen because it is expected to be sufficient.

Positioning of Weeding

To position the actuator at the location indicated by the detection system dead reckoning is going to be used. A wheel with encoder, giving a precise distance measurement, will be available already because it is also needed for the pattern recognition system.

Intra-row Weeding

Intra-row weeding will be performed by a mechanical actuator. It is expected to be difficult to remove weeds growing close to a crop plant by air, flaming, electricity, hot water, freezing, microwaves or infrared without damaging the crop plants. In that respect laser would be an excellent solution. However, laser can not work under the ground surface, and has therefore less effect on certain weed species. On the other hand, not moving the soil prevents buried seeds from germinating. A greater disadvantage of laser is its high price. High power laser is needed to reach reasonable performance, and this involves high costs. Water-jet could also probably be a good solution for intra-row weeding, but this needs much more investigation than a mechanical solution.

Determine if within Field

GPS is selected to determine wether the weeding robot is within the field or not. The determination if the weeding robot is located within the field or not, needs to be guaranteed correctly at any time. A combination of vision and dead reckoning can not give this guarantee as good as a solution in which GPS is used. Dead reckoning could improve the position determination by GPS. However, if a GPS is selected with sufficient accuracy, additional dead reckoning is not needed.

Navigate along the Row

Machine vision is selected for navigation along the row. Machine vision makes it possible to navigate along the row by relative positioning to the row. Therefore the weeding robot can work in any field without requiring absolute coordinates of a path to be followed. Absolute positioning by means of GPS, possibly combined with other sensors, requires knowledge of the absolute position of crop rows in a field. Navigation along the row by relative positioning to the row could be done also using tactile, ultrasonic or optical sensors combined with dead reckoning. Tactile sensors are not going to be used because in case of sugar beet they could harm the crop. Machine vision is preferred over ultrasonic or optical sensors, because of the ability to look forward, which contributes to a more accurate control of the position of the weeding robot relative to the crop row. It is not clear wether dead reckoning could substantially contribute to the navigation accuracy feasible with machine vision.

Determine if on Headland

GPS is selected to determine if the weeding robot is located on the headland. Using GPS requires some labour for recording the border of the headlands in advance, but will result in a correct headland detection. If a high accuracy GPS is selected, accuracy does not have to be improved by dead reckoning. Tactile, ultrasonic or optical sensors in combination with dead reckoning could also be used to determine wether the robot is on the headland, by detecting the end of the row, i.e., if over some predefined distance no row is detected. However, another crop may grow on the headland (seeded to prevent germinating of weeds) or crop rows seeded at the headland can cross the crop row to be followed. In these situations the latter solutions can not guarantee a correct detection of the end of row, and therefore also not a correct headland detection. Machine vision could give more reliable results, but it is still difficult because headland to be detected is not so structured.

Navigate on Headland

For navigation on the headland GPS is selected. On the headland the weeding robot has to make a turn and position itself in front of the next rows to be weeded. At the moment the robot arrives at the headland, a virtual path is planned to a position in front of the next rows to be weeded. Navigating over this path is going to be done by GPS.

Locomotion Related Functions

A diesel engine with a hydraulic transmission was selected for the locomotion related functions. For weeding quite some power could be required and the available power should not be limiting for realizing the objective of autonomous weeding of a field. A diesel engine with an hydraulic transmission is a proven concept in agriculture. A gearbox limits the possible combinations of the number of engine revolutions and driving speed and shuffling is difficult to automate. A continuously variable or hydraulic transmission is therefore preferred over a gearbox. Hydraulics makes it possible to design a compact wheel construction preventing damage to the crop.

A design with four wheels is preferred over one with three because of stability.

It was decided that four wheels is also preferred over two or four tracks. The most important advantages of tracks in practice are the better traction and the less soil compaction. But it is expected that if four wheels are used for such a light-weight vehicle (not more than 1500 kg) soil compaction will be acceptable. Traction when using wheels is expected to be good enough because of the limited weight and the limited need of traction for intra-row weeding. Four wheel drive and four wheel steering were chosen to have the possibility to investigate all kinds of driving strategies.

Communication with the User

Specific settings for a field will be defined by a board computer. Any moment a user wants to know the status of the weeding robot, the weeding robot status will be accessible via the internet. A website gives good opportunities to represent information in an orderly way and it is easily accessible from everywhere. In case the weeding robot needs help from its user, the weeding robot notifies its user by sending an SMS (Short Message Service) message by the GSM network. In the Netherlands any place is covered by the GSM network. From the alternatives listed, SMS is the solution that gives the highest assurance that the user really receives the message shortly after it is sent.

Detect Unsafe Situations

Detecting unsafe situations will be done super canopy all around the weeding robot. Situations in which this solution is not sufficient are hardly imaginable. Ideally the weeding robot should detect every unsafe situation, at every level and direction. Even if somebody is lying in between the crop rows below canopy level this should be detected. Because of the research effort involved in reaching the ideal objective mentioned and the possible high costs for such a solution, detecting around and only super canopy is preferred.

Fruit Picking Robots

The robotics designers offer to the farmers the opportunity to significantly reduce the costs of manual labor for harvesting. The robots can replace the seasonal manual work or even permanent employees on farms.

The robot uses computer vision algorithms to identify and locate apples in the tree. The technology used is not specifically designed for agriculture. The same technology can be applied in a wide range of industries, but for now they are using it into the agriculture.

Apples require attention at harvesting. The robot is designed to work with precision in harvesting and to store the apples. The collection is made through a flexible hose and the storage is made in the same big boxes as used by the human workers.

Autonomous Robot to Harvests Soft Fruits

Autonomous robot to harvests soft fruits.

Strawberries are delicate fruits and require careful picking.

The robot uses machine vision and motion planning algorithms to recognize and locate the ripe fruit to be picked.

The next step in robot development is to improve the learning algorithms. In order to make better fruit harvesting decisions, the robot needs algorithms capable of learning so that harvesting to be done with fewer errors.

The fruit images are captured by several cameras that move up and down for a detailed view of the crops.

The GPS system mounted on the robot platform helps to precisely establish plants and fruit production. Thus, the farmer can identify the most productive area as well as the area with a low yield.

Robotic Fruit Harvester

Robotic fruit harvester.

The robot is called FFRobot.

Like many other robots, FFrobot uses advanced image processing algorithms and algorithms for picking the fruits from the tree.

The image processing algorithms can detect damaged, diseased or unripe fruits.

The grasping hand can be easily modified to pick different types of fruits. This makes the robot available for different harvesting seasons.

SW 6010 to Harvest Strawberries

SW 6010 to harvest strawberries.

SW 6010 is the first autonomous robot available on the market able to detect and collect strawberries.

The robot has attached a number of mobile arms that can identify and pick up the ripe strawberries.

The robot analyzes the strawberries one by one and gently harvests them to prevent damage to the fruit.

AGvision is a system developed by the company and uses artificial intelligence to identify the fruits and their quality.

It can autonomously navigate between the rows of strawberries, so the operator can handle strawberries into boxes.

Multisensory System for Fruit Harvesting Robots

All harvesting robots require a sensory system that provides reliable data that can be processed and analysed in order to detect the presence of fruits, discriminate them from the rest of the scene elements and locate them spatially. In addition to complying with these fundamental objectives necessary for the efficient performance of the harvesting robot, the sensory system proposed in this study also intends to offer modularity, versatility and adaptability, so that the same rig can be utilised in various settings and with different types of crops without requiring major modifications.

The proposed multisensory system consists of an AVT Prosilica GC2450 high resolution CCD colour camera, a multispectral imaging system and a Mesa SwissRanger SR-400011 TOF 3D camera. The 5-megapixel GC2450 has a frame rate of up to 15 fps at 2448 × 2050 pixels resolution. Meanwhile, the TOF camera provides a depth map and amplitude image at the resolution of 176 × 144 pixels with 16 bit floating-point precision, as well as x, y and z coordinates to each pixel in the depth map. The detection range (radial distances) of this device goes from 0.1 m to 5.0 m, and its field of view is 69° (h) × 56° (v). The high resolution colour camera is not only utilised for the acquisition of RGB images, but also as part of the multispectral system, in which case it is set in the monochrome mode. The multispectral system is completed with a custom-made filter wheel and a servomotor that is responsible for the accurate positioning of the filter wheel. This positioning can be achieved with a maximum angular velocity of 40 rpm and a position error if 0.001°. The filter wheel allows interchanging up to five optical filters, facilitating the adaptation of the system for the detection of different kinds of crops. Since correct illumination could be critical in some scenarios, the system also includes two different light sources, an array of xenon lamps and two halogen spots, located above and at both sides of the sensory system, respectively. This lighting system is connected to a control unit that enables the independent power on and off of the lamps, and the control of their intensities. Some views of the proposed system are shown in Figure.

The RGB camera and multispectral imaging system will provide the input data required for the detection and characterisation of areas of interest that could belong to fruits, whereas the TOF 3D camera will supply simultaneously fast acquisition of accurate distances and intensity images of targets, enabling the localisation of fruits in the coordinate space. Intrinsic and extrinsic calibration parameters of both cameras were estimated by using the Matlab camera calibration toolbox.

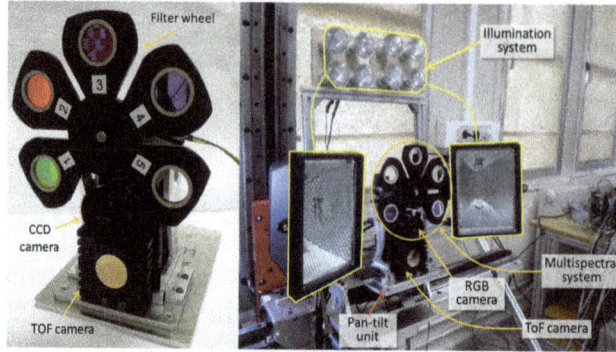

Close-up views of the multisensory system for fruit harvesting robots.

In order to confer versatility to the set-up, the whole proposed multisensory system is installed on a pan-tilt unit that facilitates the data acquisition of different viewpoints. The tilt movement has a limited angular displacement of $\alpha = \pm 30°$ relative to the horizontal axis due to mechanical constraints. The yaw movement has no mechanical constraint, so it could rotate 360° around the vertical axis. However, for the stated application, the automatic yaw movement will be restricted for azimuthal angles within the range given by $0° \leq \beta \leq 180°$.

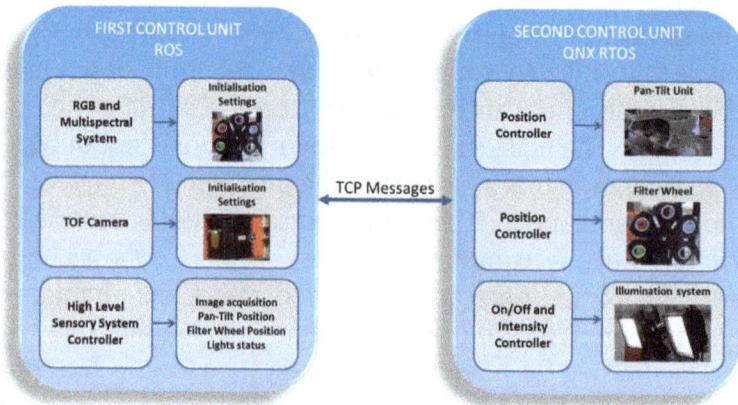

Multisensory system architecture.

The control architecture for the proposed multisensory system consists of two main parts, a unit implemented in Robot Operating System, responsible for managing the sensing devices and the high level control of the hardware elements, and a second unit implemented in QNX RTOS for the low level control of the hardware elements, which are the motorised filter wheel, the illumination system and the pan-tilt unit. Thus, the principal functions of the first unit are the initialisation of the CCD and TOF cameras (acquisition mode, pixel format), the setting of the camera parameters according to the working conditions (exposure time in the CCD camera and integration time in the TOF camera) and the control of the image acquisition procedure. Three ROS nodes are developed for achieving these functionalities: one for each camera and the sensory system controller node. Synchronous acquisition of the CDD and TOF camera is achieved when the sensory system controller publishes a trigger message that is sent when the filter wheel reaches a requested position. Immediately after the frame data acquisition is successfully completed, the sensory system controller node sends a command to the second unit implemented in QNX in order to initiate the motion of the filter wheel to the next target position. This node also sends commands for controlling the lights and the pan-tilt unit when required.

The second unit is in charge of the low level control for the high accurate positioning of the filter wheel (with a position error of $\pm 0.01285°$ and a maximum time delay of 50 ms for the positioning of each filter), switch on, switch off and intensity variation of the illumination system, as well as the high accurate positioning of the pan-tilt unit, being the PID controller the preferred option for this purpose. First and second unit communicate between them via TCP messages. These messages contain the parameters and commands required for controlling and monitoring the motion and the data acquisition tasks of the multisensory system.

Pre-processing Algorithms

Before investigating methodologies and techniques that permit us to detect and locate fruits with high accuracy, it is necessary to count with appropriate pre-processing algorithms that allow us to take full advantage of the data acquired with the designed multisensory system. Taking into consideration the configuration described in the previous subsection, two complementary pre-processing algorithms are proposed: a pixel-based classification algorithm that labels areas of interest that are candidates for belonging to fruits and a registration algorithm that combines the results of the aforementioned classification algorithm with the data provided by the TOF camera for the 3D reconstruction of the desired regions.

Several studies have demonstrated that different targets with a similar appearance when they are captured by an RGB camera can exhibit distinctive properties if they are examined with spectral systems capable of acquiring several separated wavelengths. For this reason, the first algorithm deals with the combination of RGB and filtered images acquired with the proposed multisensory system in order to achieve a classification system capable of distinguishing the different elements of the scene. The algorithm, based on Support Vector Machines (SVMs), is capable of labelling each pixel of the image into four classes that are: Stems and branches, fruits, leaves, and background. SVM is a supervised learning method utilized for classifying set of samples into two disjoint classes, which are separated by a hyperplane defined on the basis of the information contained in a training set. In the case at hand, four SVMs are utilized sequentially, each one for detecting a class against the rest. Therefore, after the first SVM is applied, pixels identified as belonging to fruit class are labelled and a mask is generated in such a way that only the remaining pixels are considered for the following SVMs. This step is then repeated for the rest of the classes in the following order: leaves, stems and branches, and finally background. The SVM classifiers are trained by selecting a random subset of samples from the RGB and filtered images and manually labelling the regions of interest from these images into the four semantic classes. The algorithm was implemented in C++ with the aid of the Open Source Computer Vision Library (OpenCV).

Once regions of interest have been detected in the scene, it is necessary to locate them spatially. The TOF camera included in the proposed multisensory system provides amplitude, depth and confidence data simultaneously for each pixel of the image captured. The amplitude represents the greyscale information, the depth is the distance value calculated within the camera and the confidence is the strength of the reflected signal, which means the quality of the depth measurements. Although TOF data is fundamental for localisation purposes, it is still necessary to automatically match this information with the classification map obtained from the previous step in a common reference frame. For accomplishing this procedure it should be taken into account that TOF images and resulting classification maps come from sensors that exhibit different field of view and different pixel array size. Thus, data will only depict the same content partially, and the pixel correspondence will not be direct. To overcome this problem, the random sample consensus (RANSAC) algorithm is adopted for the multisensory registration. RANSAC is one of the most robust algorithms for model fitting to data containing a significant percentage of errors. This iterative method estimates parameters of a mathematical model from a set of observed data which contains outliers. As the multisensory system has been designed in an enclosure that prevents the relative movements between the different elements that compose it, the idea is to use the RANSAC method to find the rotation and translation (R, T) that enable the transformation of the TOF data into the reference frame of the classification map. For that, N pairs of control point matches between Frames F_1 and F_2 are selected, where F_1 and F_2 correspond to TOF and RGB frames respectively. Note that the RGB frame is utilised for convenience, as it is consistent with that of the classification map. The control points are represented by 2D coordinates $\left(X_1^i \ \ X_2^i \right)$ in their respective reference systems. RANSAC samples the solution space of (R, T) and estimates its fitness by counting the number of inliers, f_0:

$$f_0\left(F_1, F_2, R, T\right) = \sum_i^N L\left(X_1^i, X_2^i, R, T\right)$$

where:

$$L\left(X_1^i, X_2^i, R, T\right) = \begin{cases} 1, & e = \left\| RX_1^i + T - X_2^i \right\| < \varepsilon \\ 0, & \text{otherwise} \end{cases},$$

and ε is the threshold beneath which a features match $\left(X_1^i \ \ X_2^i \right)$ is determined to be an inlier. RANSAC chooses the transform with the largest number of inlier matches. In this way, the transformation given by (R, T) may be applied to any image acquired with the TOF camera, obtaining quickly and efficiently the registered data and it won't be necessary to recalculate this transformation as long as the multisensory rig is not modified. The algorithm for on-line registration of the TOF data with the classification map was implemented in C++. Figure summarises the inputs and outputs of the proposed pre-processing algorithms.

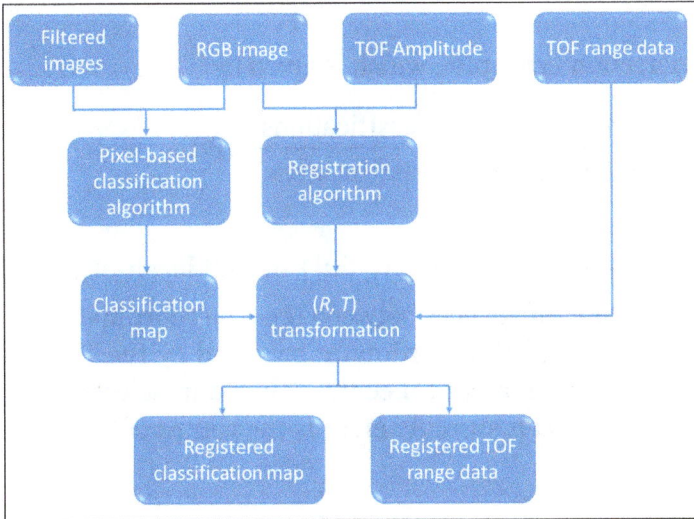

Inputs and outputs of the proposed pre-processing algorithms.

Validation Strategy

The objective of the validation strategy is to establish a structured procedure that provides quantitative information for evaluating the system performance. As it was stated before, harvesting robots require sensory systems that allow reliable detection and localisation of fruits. Thus, the quality of the proposed multisensory system and the associated set of pre-processing algorithms will be rated by comparing the obtained detection and localisation results with ground truth data that will serve as reference. The performance metrics selected for validation purposes are:

- The true positive fruit detection rate, which is a measure of the proportion of the pixels that are correctly identified as belonging to the class fruits. It is defined by:

$$TP = \frac{\text{number of pixels of the class fruits correctly classified}}{\text{total number of pixels of the class fruits}} \cdot 100\%$$

- The false positive fruit detection rate, which is the proportion of pixels that are incorrectly classified as belonging to the class fruits. It is calculated as follows:

$$FP = \frac{\text{number of pixels incorrectly classified}}{\text{total number of pixels of other classes different to fruits}} \cdot 100\%$$

- The precision of fruit detection, which is a measure of the accuracy. It is defined by:

$$Precision = \frac{TP}{TP + FP} \cdot 100\%$$

- The fruit detection error rate, which is given by:

$$Error\ rate = \frac{sum\ of\ incorrect\ classifications}{total\ number\ of\ classifications} \cdot 100\%$$

- The mean absolute error $\left[|ex|, |ey|, |ez| \right]$ in fruit localisation, which is the average of the absolute differences between the true coordinates of a selected point on the target fruit and the coordinates provided by the TOF camera, both relative to the TOF camera optical centre. The point selected on the fruit for the calculation of the mean absolute error is the centre of the visible outer surface of the fruit.

All these metrics include a clear statement of the end results expected. On the other hand, for the calculation of these performance indicators, two ground truth datasets are required. The first one should contain a list of detectable fruits, as well as their corresponding spatial localisations estimated on the centre of their visible outer surfaces. This first dataset is generated manually by one person. Thus, immediately after the data acquisition and processing of a scene, a human observer, situated in front of the crop, enumerates the visible fruits of the scene and measures their positions in the TOF camera reference frame. In this way, ground truth data generation is conducted under the same practical conditions that the data acquisition and processing. This first ground truth dataset is utilised for estimation of the mean absolute errors in order to evaluate the location capabilities of the proposed system. The second dataset includes a pixel-based masked image for each scene. The masking is performed manually on each RGB image by marking only those pixels that belong to fruits. Thus, second dataset is generated from the images acquired and processed during the experimental tests, and is used for the calculation of the rest of performance indicators with the aim of validating the detection capabilities of the proposed system.

Indoor Agriculture Robotics

Greenhouse Material Handling

With robots working safely alongside people robots can perform dreaded manual labor tasks like plant spacing. The robots can also increase plant quality by optimizing plant placement and reducing non-labor production costs including a reduction in the amount of water, pesticides, herbicides and fertilizers used as a result of more consistent spacing. Using machine vision tools and robotics new harvesting systems are starting to emerge for high value crops. The harvesting robots are able to go down greenhouse aisles accurately identify ripe versus unripe plants, harvest them and place them in on-board boxing systems.

Moving plant material is time consuming, heavy work, and expensive in terms of labor costs. There are many plant handling systems that can be used to reduce this task and its cost. The following are some typical plant handling systems used in greenhouses to improve productivity and efficiency.

Conveyors

Moving materials is one of the most labor-intensive jobs in the production of plants. Although carts are becoming more popular as a method of moving container plants, conveyors are more versatile in that it can quickly move cartons, bags, bales, bulk soil as well as plants. Conveyors are available in any width or length needed. They can be installed permanently or set up temporarily. The conveyors are installed inline so that product can be moved from one conveyor segment to the next. A conveyor should be "capacity matched" to the input rate of your materials flow. If it is not, it may cause a bottleneck, limiting the efficiency of your system. It also has to be installed and maintained properly. Many types of conveyors are used in greenhouse operations which includes belt, gravity, chain, auger, overhead monorail, etc.

Belt Conveyors

A belt conveyor works well for moving boxes, bags, bales, pots, flats and bulk materials. It may have legs or may be designed to be placed directly on the floor. If the conveyor is to be moved frequently, a carriage with pneumatic tires and a winch will make the job easier. Most manufacturers have several styles, including trough, flight, and flat belt conveyors. For most horticultural uses, a light- to medium-duty unit will provide good service and is available in widths from 4 to 24 inches and lengths to 200 feet. A lightweight, self-contained unit with an enclosed bottom works well for loading a truck or moving a shipment of containers into an elevated storage area. These units will support 300 to 400 pounds easily and are powered by an electric motor.

Gravity Conveyors

Gravity conveyors are ideal for moving boxes, bales, and flats, either by manually by workers pushing the material down the line or by allowing the maerial to move by gravity. Because of the bumpy ride the rollers create, this conveyor is better suited for products with long, flat bottoms, such as flats of boxes, rather than individual pots. There are two basic types of gravity conveyors: wheel and roller. The rolling surface of a wheel conveyor is a series of small-diameter skate wheels supported by a metal frame. Light loads and items with solid bottoms work best on this conveyor.

Chain Conveyors

Chain conveyors are used for moving of materials horizontally or between floors. They are also used in potting machines and bale breakers to move the soil vertically. The simplest system uses a single endless chain to carry pots from a potting machine to carts or trailers for transferring to the growing area. Motorized conveyors are available, but these are more expensive.

Auger Conveyors

Auger conveyors are used to convey or elevate granular materials, such as growing mixes, peat, vermiculite, sawdust, or chips. Augers are preferred for these applications to other types of conveyors because of their simplicity, low cost, durability, and versatility

Overhead Monorail Conveyors

An overhead monorail conveyor can be used to eliminate the time-consuming and backbreaking job of moving plants from the transplanting area to the growing area and on to the shipping area. This system consists of tubular or angle iron track suspended from the greenhouse frame and a trolley-mounted rack that is pushed along manually.

Gantry Conveyors

A gantry is a beam or frame, with wheels on both ends, that is supported on rails. It spans over a bay of plants and is pushed by hand. The rails, usually pipe or angle iron, are attached to the sidewall posts and extend the length of the bay.

Carts

All sizes of greenhouse operations can benefit from carts. They increase labor efficiency by allowing one person to move more material at one time than could be carried by hand. Most carts have variable shelf spacing to handle different sized plants and containers. Typically, a cart for moving bedding plants will hold 25 to 30 flats.

Forklifts

Forklifts or lift trucks are efficient in moving mate rial, but their main purpose is for lifting and stacking. Ubiquitous in warehousing operations, forklifts are commonly used in greenhouses, as warehousing or similar operations often occur here to some degree to receive, store, and distribute the supplies needed for plant production, harvesting, and transport. Most greenhouses use some degree of palletization.

Front-end Loaders

Front-end, or bucket, loaders come in three basic types: Skid-steer, wheel, and articulated-body loaders. The skid-steer loaders tend to be smaller but are used to take advantage of their size. The skid-steer method of turning enables the vehicle to turn within its own length, an advantage in tight spaces. They can also be used for other tasks.

Tractors, Trailers and other Vehicles

Although many more methods are possible for moving plants in a greenhouse, the tractorand-trailer method is a widely used one. Farm tractors a recommonly preferred as the motorized unit because they are widely available, but many other types of vehicles are used to pull trailers.

Harvest Automation Robot

The robots work safely alongside humans and require minimal training to operate, while vastly reducing production costs and improving productivity. They will reduce direct labor headcount and costs while enabling efficiency initiatives such as resource management, just in time production, and inventory control. The principal features and functions offered in the first product release include:

Features:

- Effective in greenhouse, hoop house and nursery environments.

- High placement accuracy.

- Handles most common container sizes.

- Quick swap rechargeable battery.

- Minimal training and setup.

- No programming required.

- All weather 24 hour operation.

A new approach brings a change in a variety of traditionally manual labor tasks based on mobile robot technology. Teams of small, highly intelligent machines work safely with laborers to perform the most physically demanding parts of these tasks, and at a significantly reduced cost.

Behavior based Robots

Intelligent behavior-based approach to automation provides a scalable and robust system architecture for robots that enables them to operate in even the most challenging environments. highly adaptable, behavior-based platform responds immediately to changes in the work environment, intelligently accounts for imperfect sensory data, and requires little setup and no programming.

Sensor Technology

Robots incorporate simple, highly reliable local sensing technologies that do not depend on a global model of their environment. This translates into:

- Operation not constrained by gps coverage.

- Wide variety of terrain operation.

- Ability to adapt to local conditions.

System Design and Architecture

Careful engineering and attention to our customers needs in the development stage has produced a versatile, rugged solution designed to work hard in real world environments. approach has produced the optimal result for our customers by providing equipment that is:

- Flexible and scalable.

- Fault tolerant.

- Inherently safe for operation near people.

- Easy and intuitive to set up and operate.

Agricultural Drone

Drone or UAV

A UAV (Unmanned Aerial Vehicle) is a flying device that can fly a pre-set course with the help of an autopilot and GPS coordinates. The device also has normal radio controls; it can be piloted manually in case of a fault or dangerous situation. Sometimes the term UAV is used to refer to the complete system, including ground stations and video systems, however the term is most commonly used for model planes and helicopters with both fixed and rotary wings.

Advantage

Unmanned Aerial Vehicle offers less stressful environment, it is used for better decision making, it presents safer environment, and they can fly longer hours as long as the vehicle allows for it (no human fatigue in the plane).

There is no need for the qualified pilot to fly it, in the long run,

Unmanned Air Vehicle can stay in the air for up to 30 hours, doing the repetitive tasks, performing the precise, repetitive raster scan of the region, day-after-day, night-after-night in the complete darkness or in the fog and under computer control.

Unmanned Air Vehicle performs the geological survey, it performs the visual or thermal imaging of the region, it can measure the cell phone, radio or TV coverage over any terrain, the drone pilots or operators can easily hand off controls of the drone without any operational downtime.

The drones can have more pinpoint accuracy from greater distances.

Basic Principle – Working of Drone

The 4 propellers of a drone or quadcopter are fixed and vertically orientated. Each propeller has a variable and independent speed which allows a full range of movements. The core components of a drone are as follows:

Chassis: The skeleton of the drone which all componentry is fixed to. The chassis design is a trade-off between strength (especially when additional weights such as cameras are attached) and additional weight, which will require longer propellers and stronger motors to lift.

Propellers: Principally effect load the drone can carry the speed it can fly and the speed it can manoeuvre. The length can be modified; longer propellers can achieve greater lift at a lower rpm but take longer to speed up/slow down. Shorter propellers can change speed quicker and thus are more manoeuvrable; however they require a higher rotational speed to achieve the same power as longer blades. This causes excess motor strain and thus reduces motor life span. A more aggressive pitch will allow quicker movement but reduced hovering efficiency.

Motors: 1 per propeller, drone motors are rated in "kV" units which equates to the number of revolutions per minute it can achieve when a voltage of 1 volt is supplied to the motor with no load. A faster motor spin will give more flight power, but requires more power from the battery resulting in a decreased flight time.

Electronic Speed Controller (ESC): Provides a controlled current to each motor to produce the correct spin speed and direction.

Flight Controller: The onboard computer which interprets incoming signals sent from the pilot and sends corresponding inputs to the ESC to control the quadcopter.

Radio Receiver: Receives the control signals from the pilot.

Battery: Generally lithium polymer batteries are used due to high power density and ability to recharge.

Further to this, sensors can be used such as accelerometers, gyroscopes, GPS and barometers for positional measurements. Cameras are also frequently mounted for navigation and aerial photography.

Drone Mechanism – Flying a Quadcopter Drone

A drone is controlled manually with a hand- held radio control transmitter which manually controls the propellers. Sticks on the controller allow movements in different directions and trim buttons allow the trim to be adjusted to balance the drone. Screens can also be used to receive live video footage from the on-board camera and to display sensor data.

Further to this, on-board sensors can provide helpful settings such as;

- Auto altitude where the drone will move at a fixed altitude,
- GPS hold, where the drone will remain at a fixed GPS position.

Drone can also be flown autonomously, modern flight controllers can use software to mark GPS waypoints that the vehicle will fly to and land or move to a set altitude.

Uses of Agricultural Drones

Agricultural drones help to achieve and improve what's known as precision agriculture.

This approach to farming management is based on observing, measuring, and taking action based on real-time crop and livestock data. It erases the need for guesswork in modern farming and instead gives farmers the ability to maximize their yields and run more efficient organizations, all while enhancing crop production.

In recent years the cost of agriculture drones has rapidly declined, which has not only led to the explosion of drone use cases in agriculture but has made it a no-brainer investment for modern farmers.

In fact, the agricultural drone market is expected to grow over 38% in coming years. Driven by growing population levels and changing climate patterns, the need for efficient agriculture is only going to become more important.

There are multiple uses for agricultural drones, including:

- Scouting land and crops,

- Checking for weeds and spot treating plants,

- Monitoring overall crop health,

- Managing livestock and monitoring for health issues,

- and more.

Drones are equipped with technology like propulsion systems, infrared cameras, GPS and navigation systems, programmable controllers, and automated flight planning. Plus, with custom-made data processing software any collected information can instantly be put to use towards better management decisions.

Drones and Agriculture: A Match Made in Heaven

Drones are transforming how agriculture and farming are done.

By implementing drone technology, farms and agriculture businesses can improve crop yields, save time, and make land management decisions that'll improve long-term success.

Farmer's today have a variety of complex factors that influence the success of their farms. From water access to changing climate, wind, soil quality, the presence of weeds and insects, variable growing seasons, and more.

As a result, farmers are turning to high-level drone technology to help remedy these problems, and provide fast and efficient solutions.

Agricultural drones allow farmers to obtain access to a wealth of data they can use to make better management decisions, improve crop yields, and increase overall profitability.

Drones can be used to collect data related to crop yields, livestock health, soil quality, nutrient measurements, weather and rainfall results, and more. This data can then be used to get a more accurate map of any existing issues, as well as create solutions based upon extremely reliable data.

The agriculture industry is no stranger to embracing changing technological trends to streamline business. The use of drones in agriculture is the next technological wave that'll help agricultural businesses meet the changing and growing demands of the future.

How Agricultural Drones are used: 6 Innovative Methods

The use of drones in agriculture is here to stay.

Drone technology can help to accomplish once time-consuming and difficult tasks, all while reducing costs across the board.

You can expect the current uses of drones in agriculture to continue to evolve as the industry matures and new technology is introduced.

Currently, there are six common uses of agricultural drones, which we profile below:

Soil and Field Analysis

At the beginning, middle, and end of a crop cycle drones can be used to help obtain useful data surrounding the quality of the existing soil. By obtaining 3D maps of existing soil, you'll be able to see if there are any issues surrounding soil quality, nutrient management, or soil dead zones.

This information can help farmers determine the most effective patterns for planting, managing crops, soil, and more. Ongoing monitoring can help to better utilize water resources, and more effectively manage crop nutrient levels.

Seed Planting

Drone planting is a relatively newer technology and not as widely used, but some companies are experimenting with drone planting. Essentially, manufacturers are experimenting with custom systems that have the ability to shoot seed pods into prepared soil.

Drone startup companies have been instrumental in developing unique drone technologies to assist with a wide range of ecological and agricultural issues. For example, the company DroneSeed is using unmanned aircraft capable of delivering up to 57 pounds of payload in the form of tree seeds, herbicides, fertilizer and water per aircraft per flight to assist reforestation and replanting projects.

This technology helps to minimize the need for on-the-ground planting, which can be costly, time-intensive, and strenuous work.

This same drone technology can be adapted and applied to a wide range of farm types, reducing overall planting times and labor costs across the board.

Crop Spraying and Spot Spraying

Crops require consistent fertilization and spraying in order to maintain high yields. Traditionally this was done manually, with vehicles, or even via airplane. These methods are not only inefficient, and burdensome, but they can be very costly as well.

With approval from the FAA, Drones can be equipped with large reservoirs, which can be filled with fertilizers, herbicides, or pesticides. Using drones for crop spraying is much safer and cost-effective. Drones can even be operated completely autonomously and programmed to run on specific schedules and routes.

For example, if there's a fungus breakout in a certain section of the crops, drones can be used to spot treat the issue. With the speed at which drones can operate, you can diagnose and treat potential crop issues before they become a widespread issue across the entire farm.

Spot spraying of crops used to be incredibly difficult. If you had an issue with weeds or a certain crop, the entire acreage would have to be sprayed.

This is a huge waste of time and resources, as someone will have to walk the entire acreage, plus there are the overall costs of pesticides and the associated environmental cost of chemical usage.

With spot spraying afforded by drones, this same task can be accomplished in less time, with fewer monetary resources, and a reduced environmental cost.

Crop Mapping and Surveying

One of the biggest advantages of using drone technology is the ease and effectiveness of large-scale crop and acreage monitoring. In the past, satellite or plane imagery was used to help get a large scale view of the farm, while helping to spot potential issues.

However, these images were not only expensive but lacked the precision that drones can provide. Today, you can not only obtain real-time footage but also time-based animation which can illuminate crop progression in real-time.

With drone mapping and surveying, technology decisions can now be made based on real-time data, not outdated imagery, or best-practice guesswork.

With near infrared (NIR) drone sensors you can actually determine plant health based upon light absorption, giving you a birds-eye view of the overall farm health. We recently interviewed a drone pilot who used NIR to help vineyard owners determine the health of their grapevines.

With agriculture drones you'll be able to collect information like:

- The overall crop and plant health.
- Land distribution based on crop type.
- Current crop life cycle.
- Detailed GPS maps of current crop area.

The end result is simple, drones can help to maximize land and resource usage, and help farmers better determine crop planting locations.

Irrigation Monitoring and Management

Irrigation can be troublesome. With miles and miles of irrigation, issues are bound to arise. Drones that are equipped with thermal cameras can help to spot irrigation issues, or areas that are receiving too little or excessive moisture.

With this information, crops can be better laid out to maximize drainage, adhere to natural land runoff, and avoid water pooling, which can damage sensitive crops.

Water and irrigation issues are not only costly but can ruin crop yields as well. With drone surveying, these issues can be spotted before they become troublesome.

Real-time Livestock Monitoring

Some drones are equipped with thermal imaging cameras that enable a single pilot to manage and monitor livestock. This allows farmers to keep track of livestock a much greater frequency, and with less time and staff investment.

The drone operator can quickly check in on herd to see if there are any injured or missing livestock, as well as see livestock who are giving birth. Drones are used to keep an eye on the herd at all times, a once costly and time-intensive task.

Plus, thermal imaging will also help to keep an eye out for any livestock predators, which can be a huge advantage for some farm owners.

Hydraulic Actuators

Pneumatic actuators are normally used to control processes requiring quick and accurate response, as they do not require a large amount of motive force.

However, when a large amount of force is required to operate a valve (for example, the main steam system valves), hydraulic actuators are normally used.

Although hydraulic actuators come in many designs, piston types are most common.

A typical piston-type hydraulic actuator is shown in Below Figure. It consists of a cylinder, piston, spring, hydraulic supply and return line, and stem.

The piston slides vertically inside the cylinder and separates the cylinder into two chambers. The upper chamber contains the spring and the lower chamber contains hydraulic oil.

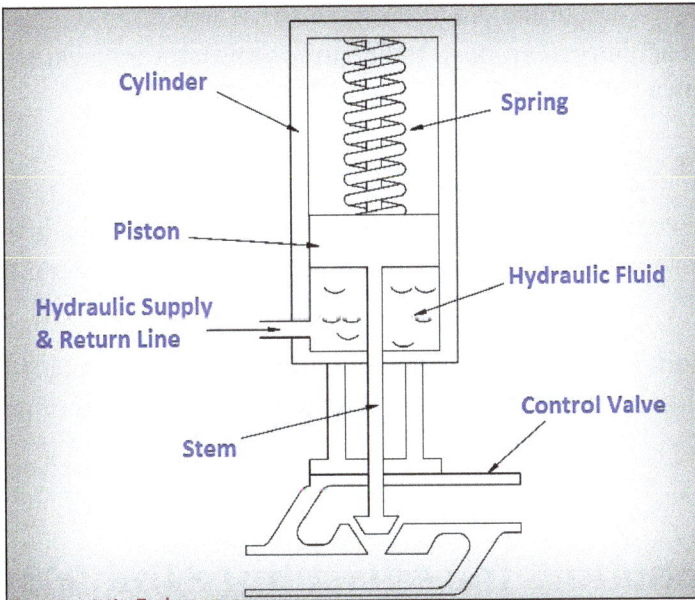

The hydraulic supply and return line is connected to the lower chamber and allows hydraulic fluid to flow to and from the lower chamber of the actuator. The stem transmits the motion of the piston to a valve.

Initially, with no hydraulic fluid pressure, the spring force holds the valve in the closed position. As fluid enters the lower chamber, pressure in the chamber increases.

This pressure results in a force on the bottom of the piston opposite to the force caused by the spring. When the hydraulic force is greater than the spring force, the piston begins to move upward, the spring compresses, and the valve begins to open.

As the hydraulic pressure increases, the valve continues to open. Conversely, as hydraulic oil is drained from the cylinder, the hydraulic force becomes less than the spring force, the piston moves downward, and the valve closes. By regulating amount of oil supplied or drained from the actuator, the valve can be positioned between fully open and fully closed.

The principles of operation of a hydraulic actuator are like those of the pneumatic actuator. Each uses some motive force to overcome spring force to move the valve. Also, hydraulic actuators can be designed to fail-open or fail-closed to provide a fail-safe feature.

Advantages of Hydraulic Actuators

- Hydraulic actuators are rugged and suited for high force applications. They can produce forces 25 times greater than pneumatic cylinders of equal size. They also operate in pressures of up to 4,000 psi.

- A hydraulic actuator can hold force and torque constant without the pump supplying more fluid or pressure due to the incompressibility of fluids.

- Hydraulic actuators can have their pumps and motors located a considerable distance away with minimal loss of power.

Disadvantages of Hydraulic Actuators

Hydraulics will leak fluid. Like pneumatic actuators, loss of fluid leads to less efficiency and cleanliness problems resulting in potential damage to surrounding components and areas.

Hydraulic actuators require many complementary parts, including a fluid reservoir, motor, pump, release valves, and heat exchangers, along with noise reduction equipment.

Linear Actuators for Agriculture Industry

The modern agricultural industry depends a great deal on small and large farming equipment. Many machines, such as corn detasselers, threshers, and fertilizer sprayers are automated to ensure that the farming process is performed and completed with precision. Linear actuators can support a variety of agricultural applications. They support agricultural equipment by providing precise motion control for various components, thus enabling the completion of various farming processes.

Electric Linear Actuators Support the Agricultural Industry

Farmers have a work schedule that begins before sunrise and ends around sunset. Numerous activities need to be completed within this timeframe. Therefore, the equipment need to work precisely to a timed schedule for the daily work to be completed. Also, farming equipment is utilized in extreme weather, which can lead to problems like corrosion, and dirt and dust accumulation. Hence, linear actuators for agricultural industry needs to be designed for efficiency and durability.

The benefits of electric linear actuators can help you understand why they are the best choice against hydraulic actuators.

- Flexibility: They can be designed in a variety of sizes and power ranges. The design can be done to create seamless interactions with control systems.

- Accuracy: Many large sized equipment require forces up to 10,000 N. An electric actuator can provide the exact amount of power for an application. Feedback provided in terms of positioning is extremely accurate.

- Risk and maintenance-free: Electric linear actuators provide many years of service without requiring maintenance. They eliminate the risks related to leaking of hydraulic actuators. Also, linear actuators consume less power compared to hydraulic actuators.

Linear Actuators in Agricultural Applications

Given below are small as well as large applications, which make use of electric linear actuators.

- Fertilizer spreader: Agricultural actuators can be used to regulate the amount of fertilizer that is dispensed onto field crops. They accomplish this by way of a controlled system, allowing the farmer to calibrate the automatic spreader according to the field layout and fertilizer requirement. Potentiometers communicate with the control system so that the spreader can dispense only the right amount of fertilizer on crops.

- Combine harvesters: Electric linear actuators can provide many solutions for combine harvesters. The actuator control systems can be used to adjust the position of the cutter bar, and the speed of the knives on the bar. The covers of grain tanks can be opened and closed easily using linear motion control, allowing for easy grain removal.

- Corn headers: Corn headers are designed to provide cleaner corn loads. They consist of large teeth that tear the corn from the stalks. Linear actuators can be used to maintain the exact distance or space where the corn cobs will be cut from the cane or stalk.

- Miscellaneous applications: Other smaller sized application where linear actuators are used include:

  - Opening and closing of barn or stable doors.

  - Activation of ventilation systems of storage sheds.

  - Timely opening of feed gates for animals.

Electric linear actuators can be adapted to provide solutions for a variety of farming applications. Whether its complete motion control or precise adjustments in movement, linear actuators prove to be indispensable products for the agricultural industry.

References

- Agbot-agricultural-robot: whatis.techtarget.com, Retrieved 19 May, 2019

- 8-Types-of-Field-Robots-in-Agriculture-113: robotics.org, Retrieved 15 April, 2019

- Robotic-applications-in-agricultural-industry-autonomous-agricultural-robot-types-uses-and-importance: online-sciences.com, Retrieved 19 June, 2019

- An-Autonomous-Weeding-Robot-for-Organic-Farming- 221323085: researchgate.net, Retrieved 18 April, 2019

- Fruit-harvesting-robots: intorobotics.com, Retrieved 16 July, 2019

- Agricultural-drones: uavcoach.com, Retrieved 23 May, 2019

- What-is-a-hydraulic-actuator: instrumentationtools.com, Retrieved 08 January, 2019

- Linear-actuators-for-agriculture-industry-achieving-precise-motion-control: venturemfgco.com, Retrieved 14 February, 2019

Automation and IoT in Agriculture

Automation and IoT are used in different agricultural purposes such as weed control, cloud seeding, planting seeds, harvesting, environmental monitoring and soil analysis. This chapter closely examines these applications of automation and IoT in agriculture to provide an extensive understanding of the subject.

IoT Applications in Agriculture

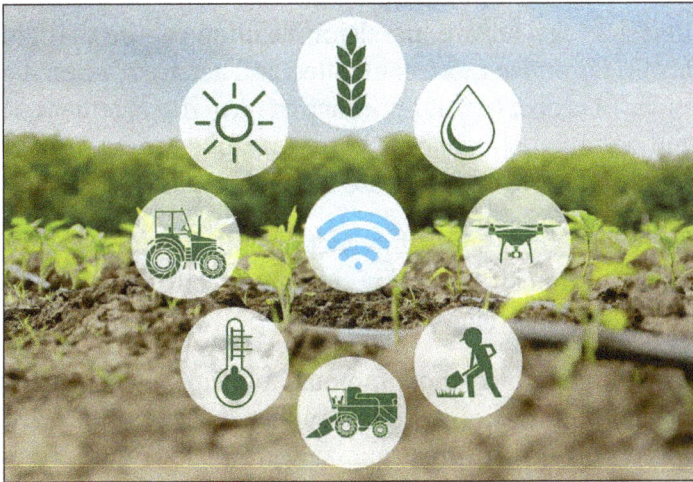

IoT has the capability to influence the world we live in; advanced industries, connected vehicles, and smarter cities are all components of the IoT equation. However, applying technology like IoT to the agriculture industry could have the greatest impact.

The global population is set to touch 9.6 billion by 2050. So, to feed this much population, the farming industry must embrace IoT. Against the challenges such as extreme weather conditions and rising climate change, and environmental impact resulting from intensive farming practices, the demand for more food has to be met.

Smart farming based on IoT technologies will enable growers and farmers to reduce waste and enhance productivity ranging from the quantity of fertilizer utilized to the number of journeys the farm vehicles have made.

In IoT-based smart farming, a system is built for monitoring the crop field with the help of sensors (light, humidity, temperature, soil moisture, etc.) and automating the irrigation system. The farmers can monitor the field conditions from anywhere. IoT-based smart farming is highly efficient when compared with the conventional approach.

The applications of IoT-based smart farming not only target conventional, large farming operations, but could also be new levers to uplift other growing or common trends in agricultural like organic farming, family farming (complex or small spaces, particular cattle and/or cultures, preservation of particular or high quality varieties etc.), and enhance highly transparent farming.

In terms of environmental issues, IoT-based smart farming can provide great benefits including more efficient water usage, or optimization of inputs and treatments.

Applications of IoT in Agriculture

Precision Farming

Also known as precision agriculture, precision farming can be thought of as anything that makes the farming practice more controlled and accurate when it comes to raising livestock and growing of crops. In this approach of farm management, a key component is the use of IT and various items like sensors, control systems, robotics, autonomous vehicles, automated hardware, variable rate technology, and so on.

The adoption of access to high-speed internet, mobile devices, and reliable, low-cost satellites (for imagery and positioning) by the manufacturer are few key technologies characterizing the precision agriculture trend.

Precision agriculture is one of the most famous applications of IoT in the agricultural sector and numerous organizations are leveraging this technique around the world. CropMetrics is a precision agriculture organization focused on ultra-modern agronomic solutions while specializing in the management of precision irrigation.

The products and services of CropMetrics include VRI optimization, soil moisture probes, virtual optimizer PRO, and so on. VRI (Variable Rate Irrigation) optimization maximizes profitability on irrigated crop fields with topography or soil variability, improve yields, and increases water use efficiency.

The soil moisture probe technology provides complete in-season local agronomy support, and recommendations to optimize water use efficiency. The virtual optimizer PRO combines various technologies for water management into one central, cloud based, and powerful location designed for consultants and growers to take advantage of the benefits in precision irrigation via a simplified interface.

Agricultural Drones

Technology has changed over time and agricultural drones are a very good example of this. Today, agriculture is one of the major industries to incorporate drones. Drones are being used in agriculture in order to enhance various agricultural practices. The ways ground-based and aerial based drones are being used in agriculture are crop health assessment, irrigation, crop monitoring, crop spraying, planting, and soil and field analysis.

The major benefits of using drones include crop health imaging, integrated GIS mapping, ease of use, saves time, and the potential to increase yields. With strategy and planning based on real-time data collection and processing, the drone technology will give a high-tech makeover to the agriculture industry.

IoT uses drones for gathering valuable data via a series of sensors that are used for imaging, mapping, and surveying of agricultural land. These drones perform in-flight monitoring and observations. The farmers enter the details of what field to survey, and select an altitude or ground resolution.

From the drone data, we can draw insights regarding plant health indices, plant counting and yield prediction, plant height measurement, canopy cover mapping, field water ponsing mapping, scouting reports, stockpile measuring, chlorophyll measurement, nitrogen content in wheat, drainage mapping, weed pressure mapping, and so on.

The drone collects multispectral, thermal, and visual imagery during the flight and then lands in the same location it took off.

Livestock Monitoring

Large farm owners can utilize wireless IoT applications to collect data regarding the location, well-being, and health of their cattle. This information helps them in identifying animals that are sick so they can be separated from the herd, thereby preventing the spread of disease. It also lowers labor costs as ranchers can locate their cattle with the help of IoT based sensors.

Many organizations offer cow monitoring solutions to cattle producers. One of the solutions helps the cattle owners observe cows that are pregnant and about to give birth. From the heifer, a sensor powered by battery is expelled when its water breaks. This sends an information to the herd manager or the rancher. In the time that is spent with heifers that are giving birth, the sensor enables farmers to be more focused.

Smart Greenhouses

Greenhouse farming is a methodology that helps in enhancing the yield of vegetables, fruits, crops etc. Greenhouses control the environmental parameters through manual intervention or a proportional control mechanism. As manual intervention results in production loss, energy loss, and labor cost, these methods are less effective. A smart greenhouse can be designed with the help of IoT; this design intelligently monitors as well as controls the climate, eliminating the need for manual intervention.

For controlling the environment in a smart greenhouse, different sensors that measure the environmental parameters according to the plant requirement are used. We can create a cloud server for remotely accessing the system when it is connected using IoT.

This eliminates the need for constant manual monitoring. Inside the greenhouse, the cloud server also enables data processing and applies a control action. This design provides cost-effective and optimal solutions to the farmers with minimal manual intervention.

Some organizations use new modern technologies for providing services. It builds modern and affordable greenhouses by using solar powered IoT sensors. With these sensors, the greenhouse state and water consumption can be monitored via SMS alerts to the farmer with an online portal. Automatic Irrigation is carried out in these greenhouses.

The IoT sensors in the greenhouse provide information on the light levels, pressure, humidity, and temperature. These sensors can control the actuators automatically to open a window, turn on lights, control a heater, turn on a mister or turn on a fan, all controlled through a WiFi signal.

It is not a secret that the Internet of Things (IoT) triumphally changes the world. In fact, it has already introduced innovation in various industries, which assisted in increasing the effectiveness and cutting the costs of business operations in different aspects. And the area of agriculture fits this trend totally. Being previously dependent on human resources and hard machinery completely, it has also started applying technological solutions and modernizing its core operations. And so, it is possible to discuss agriculture IoT as the whole sphere. To address this task, we discover the main directions in which Internet of Things (IoT) applications in agriculture managed to make a significant impact.

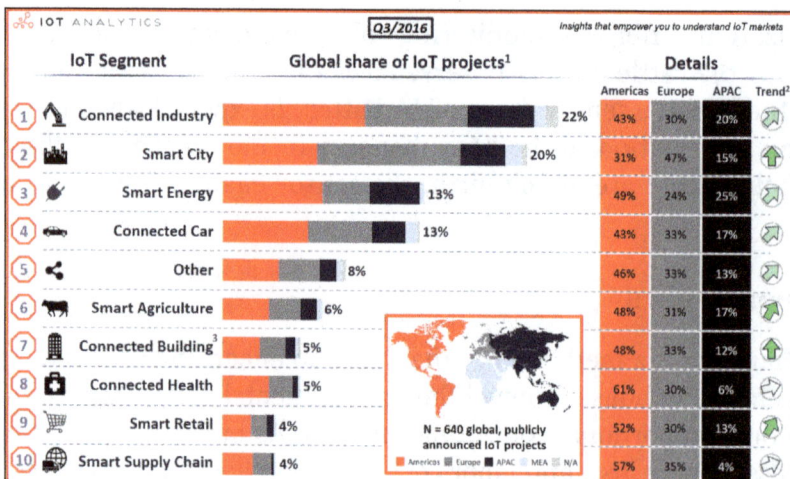

	IoT Segment	Global share of IoT projects[1]	Americas	Europe	APAC	Trend[2]
①	Connected Industry	22%	43%	30%	20%	
②	Smart City	20%	31%	47%	15%	
③	Smart Energy	13%	49%	24%	25%	
④	Connected Car	13%	43%	33%	17%	
⑤	Other	8%	46%	33%	13%	
⑥	Smart Agriculture	6%	48%	31%	17%	
⑦	Connected Building[3]	5%	48%	33%	12%	
⑧	Connected Health	5%	61%	30%	6%	
⑨	Smart Retail	4%	52%	30%	13%	
⑩	Smart Supply Chain	4%	57%	35%	4%	

Q3/2016 — Insights that empower you to understand IoT markets

N = 640 global, publicly announced IoT projects

IoT Farming: The most Widespread Direction

As one of the popular dimensions to discuss applications of IoT in agriculture, the precision farming deserves special attention. In this sphere, the adoption of smart technology includes using sensors, robots, systems of control, and autonomous vehicles. In addition, the potential of IoT farming includes an ability to offer for farmers environmentally friendly pesticides. Besides, the introduction of smart technology into agriculture enables proper tracking of the natural factors, like climate change, soil composition, and weather forecast.

In this dimension, it is possible to mention several cases of IoT applications in agriculture. Firstly, there exists the entire CropMetrics organization that works with VRI (Variable Rate Irrigation) optimization, meaning an ability to improve topography or soil variability and maximize efficiency and yield performance. In this case, IoT farming means inviting a precision data specialist who introduces simple-to-use cloud software with a high level of customization that optimizes irrigation scheduling and maximizes profit.

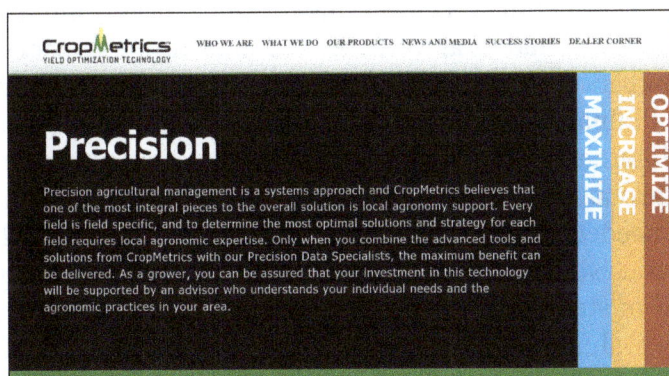

Secondly, farmers widely use Arable and Semios to monitor the state of their crops. On the one hand, Arable allows growers to use an analytics platform that provides a unique opportunity to collect both weather and plant information and integrate it into a cloud. On another hand, IoT in agriculture as a Semios solution means using a scalable platform for yield improvement with real-time updates on the health condition of plants.

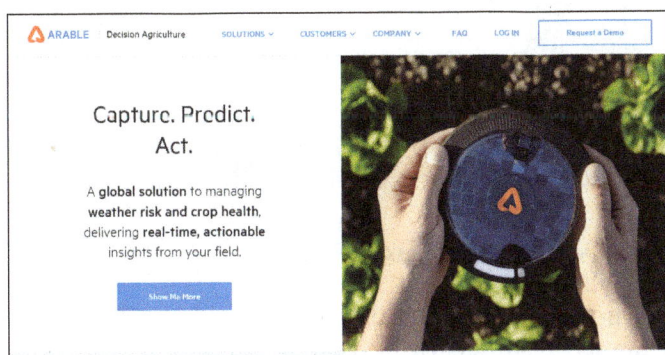

Finally, there is a line of IoT applications in smart agriculture aimed at climate change prediction. In this context, allMETEO and its Meteoshield and Smart City weather sensors allow smarter management of crops with notifications of the necessary preventive measures to protect the plants. In fact, the adoption of this innovation significantly improves the potential of precision farming too.

IoT Applications in Agriculture: The Assistance of Drones

Drones are highly useful in managing agricultural processes — with the pictures and aerial maps they provide, a farmer gets an immediate understanding of which crops need urgent attention. Besides, the advantages of using this innovation in the sphere include better care over crop in general — evaluation of its health state, irrigation, monitoring of progress, spraying, and planting. Finally, drones are helpful in saving time, since all the information is collected without the need to invest time and effort in working on the field.

In practice, IoT offers for agricultural needs two types of drones: ground-based and aerial-based ones. To gather the necessary information, farmers input the field data, including ground resolution and its altitude. As a result, a drone provides details on plant counting, yield prediction, health indices, height meterage, the presence of chemicals in plants and soil, drainage mapping and various other data. Among the examples from the sphere of the Internet of Things agriculture, the basic directions of drone assistance include soil and field analysis (with 3D maps for seed planting predictions), planting (by providing the needed nutrients), crop spraying (with ultrasonic echoing and lasers to adjust altitude and avoid collisions), crop monitoring (through providing time-series animation instead of static satellite images), irrigation (having sensors to reveal dry areas), and health evaluation (taking crop scans to identify the lack of green light and NIR light). In other words, drones take care of the full cycle of crops.

IoT Applications for Greenhouse Farming

As the special dimension of IoT applications in agriculture, there exist various effective solutions for greenhouse farming. In particular, climate control is achieved through positioning several sensors that send alerts about water or air problems. In practice, the products that allow achieving these aims include Farmapp and Growlink.

Farmapp, being a representative of IoT applications in smart agriculture, offers farmers an Integrated Pest Management software with monitoring, sensors, and fumigation functions. Specifically, it includes a scouting app for fast recording and implementation of the needed measures — with satellite maps, comparative maps, charts and reports at hand. Moreover, it is possible to receive real-time data on weather and soil condition through a direct access to satellite images and algorithmic calculations. Finally, the functionality of Farmapp captures better irrigation — in this dimension, this IoT in agriculture enables tracking of the amount of water spent on plants for its optimization.

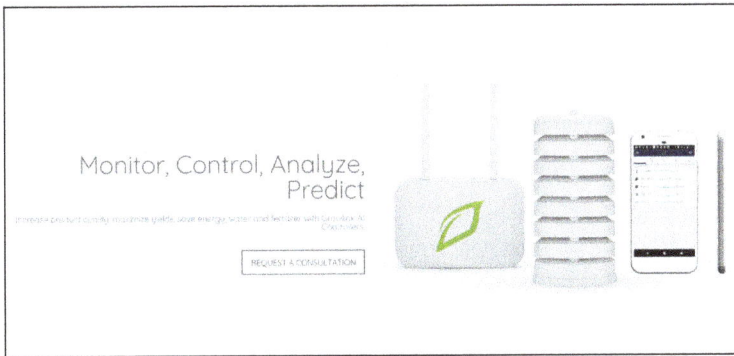

Monitor, Control, Analyze, Predict

REQUEST A CONSULTATION

In its turn, Growlink allows a real-time monitoring in greenhouses with the aim to increase quality and yield performance. In particular, this Internet of Things agriculture solution concentrates on automation of working with operational data — including planning, controlling, tracking, and monitoring activities. Hence, farmers get an outstanding opportunity to achieve the best performance possible in a long run.

Internet of Things Agriculture Solutions for Livestock Management

As for the livestock control, IoT in agriculture assists in tracking the state of the herd in general and each its representative in particular. In this sphere, there exist applications to determine the health of animals, find their location, and track the state of pregnancies — especially, while dealing with cattle and chicken. And among the concrete examples among IoT applications in agriculture in this dimension, there are SCR by Allflex and Cowlar.

SCR by Allflex offers cow, milking, and herd intelligence, along with several other professional solutions. In this context, the functionality of its service includes tracking all the insights about each herd participant (heat, health, and nutrition), optimizing milking process (simplify and streamline), and collecting data into an integrated and actionable plan for herd development. Likewise, Cowlar is a company that addresses the similar needs — optimizing milking, maximizing performance, reducing labor costs — along with boosting reproduction.

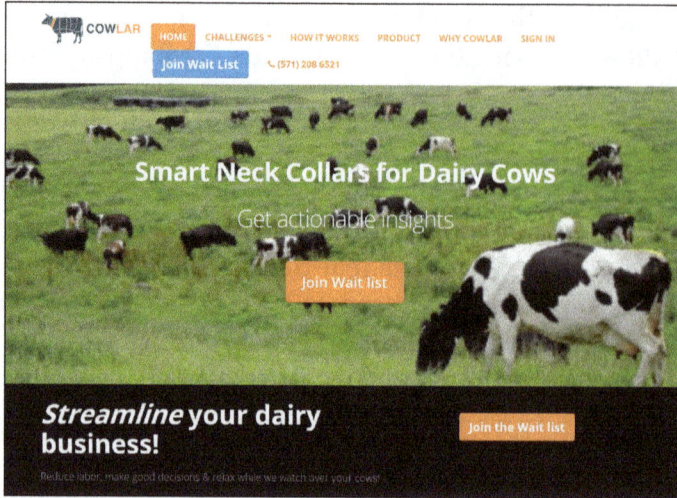

Besides, to address the challenge of the need for monitoring animal health in a long run, Symphony Link is an application that avoids mesh networking and completes the task of a complete integration effectively. As a revolutionary invention in the world of IoT in agriculture, it links wide-area IoT networks with modules (including RXR-27), gateways, and conductor.

IoT Applications for the Entire Supply Chain in Agriculture

In terms of logistics, Internet of Things agriculture enables using GPS, RFID, and other location-based sensors to control transportation and storing of plants. In this context, the entire supply chain can increase its effectiveness, meaning the improvements in terms of transparency and customer awareness (precisely, in terms of food safety).

Furthermore, end-to-end farm management systems are also an area of the interest of IoT software development in farming market. In this context, there exists a possibility to install sensors and devices that can provide data for analytics, reports, and accounting. The exact solutions with these features are FarmLogs and Cropio.

FarmLogs presents on agriculture market a software for facilitating grain marketing decisions. Specifically, it provides a toolkit necessary for creating a grain marketing plan (with the value of unsold crops, contact list, and goal-setting) and insights on profitability increasing. Among the concrete products, farmers can order marketing, reports, automatic activity recording, crop health imagery, and rainfall tracking.

As for Cropio, the solution refers to field management and vegetation control system functionality. Specifically, it facilitates checking the state of numerous fields, provides real-time data on the necessary updates, and assists in forecasting. Among its key features, the abilities to provide field history, instant alerts, vegetation map, soil moisture, and harvest forecast are impressive.

IoT is Shaping Agriculture

Technologies and IoT have the potential to transform agriculture in many aspects. Namely, there are five ways IoT can improve agriculture:

- Data, tons of data, collected by smart agriculture sensors, e.g. weather conditions, soil quality, crop's growth progress or cattle health. This data can be used to track the state of your business in general, as well as staff performance, equipment efficiency, etc.

- Better control over the internal processes and, as a result, lower production risks. The ability to foresee the output of your production allows you to plan for better product distribution. If you know exactly how much crops you are going to harvest, you can make sure your product won't lie around unsold.

- Cost management and waste reduction thanks to the increased control over production. Being able to see any anomalies in the crop growth or livestock health, you will be able to mitigate the risks of losing your yield.

- Increased business efficiency through process automation. By using smart devices, you can automate multiple processes across your production cycle, e.g. irrigation, fertilizing, or pest control.

- Enhanced product quality and volumes. Achieve better control over the production process and maintain higher standards of crop quality and growth capacity through automation.

As a result, all of these factors can eventually lead to higher revenue.

IoT use Cases in Agriculture

There are many types of IoT sensors and IoT applications that can be used in agriculture:

Monitoring of Climate Conditions

Probably the most popular smart agriculture gadgets are weather stations, combining various smart farming sensors. Located across the field, they collect various data from the environment and send it to the cloud. The provided measurements can be used to map the climate conditions, choose the appropriate crops, and take the required measures to improve their capacity (i.e. precision farming).

Some examples of such agriculture IoT devices are allMETEO, Smart Elements, and Pycno.

Greenhouse Automation

In addition to sourcing environmental data, weather stations can automatically adjust the conditions to match the given parameters. Specifically, greenhouse automation systems use a similar principle.

For instance, Farmapp and Growlink are also IoT agriculture products offering such capabilities among others.

Crop Management

One more type of IoT product in agriculture and another element of precision farming is crop management devices. Just like weather stations, they should be placed in the field to collect data specific to crop farming; from temperature and precipitation to leaf water potential and overall crop health, these can all be used to readily collect data and information for improved farming practices.

Thus, you can monitor your crop growth and any anomalies to effectively prevent diseases or infestations that could harm your yield. Arable and Semios can serve as good representations of how this use case can be applied in real life.

Cattle Monitoring and Management

Just like crop monitoring, there are IoT agriculture sensors that can be attached to the animals on a farm to monitor their health and log performance. This works similarly to IoT devices for pet care.

For example, SCR by Allflex and Cowlar use smart agriculture sensors (collar tags) to deliver temperature, health, activity, and nutrition insights on each individual cow, as well as collective information about the herd.

End-to-end Farm Management Systems

A more complex approach to IoT products in agriculture can be represented by the so-called farm productivity management systems. They usually include a number of agriculture IoT devices and sensors, installed on the premises as well as a powerful dashboard with analytical capabilities and in-built accounting/reporting features.

This offers remote farm monitoring capabilities and allows you to streamline most of the business operations. Similar solutions are represented by FarmLogs and Cropio.

In addition to the listed IoT agriculture use cases, some prominent opportunities include vehicle tracking (or even automation), storage management, logistics, etc.

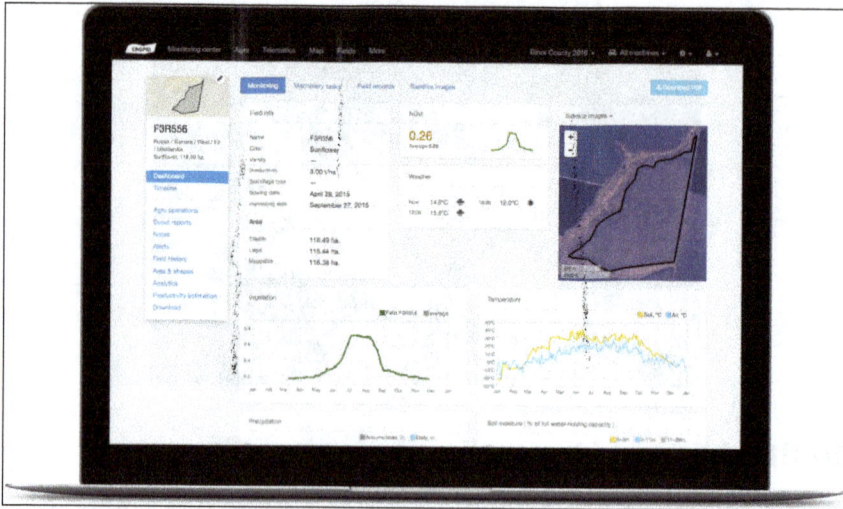

Four Things to Consider before Developing your Smart Farming Solution

As we can see, the use cases for IoT in agriculture are endless. There are many ways smart devices can help you increase your farm's performance and revenue. However, agriculture IoT apps development is no easy task. There are certain challenges you need to be aware of if you are considering investing in smart farming.

Hardware

To build an IoT solution for agriculture, you need to choose the sensors for your device (or create a custom one). Your choice will depend on the types of information you want to collect and the purpose of your solution. In any case, the quality of your sensors is crucial to the success of your product— it will depend on the accuracy of the collected data and its reliability.

The Brain

Data analytics should be at the core of every smart agriculture solution. The collected data itself will be of little help if you cannot make sense of it. Thus, you need to have powerful data analytics capabilities and apply predictive algorithms and machine learning in order to obtain actionable insights based on the collected data.

Maintenance

Maintenance of your hardware is a challenge that is of primary importance for IoT products in agriculture, as the sensors are typically used in the field and can be easily damaged. Thus, you need to make sure your hardware is durable and easy to maintain. Otherwise, you will need to replace your sensors more often than you would like.

Mobility

Smart farming applications should be tailored for use in the field. A business owner or farm manager should be able to access the information on site or remotely via a smartphone or desktop computer.

Plus, each connected device should be autonomous and have enough wireless range to communicate with the other devices and send data to the central server.

Infrastructure

To ensure that your smart farming application performs well (and to make sure it can handle the data load), you need a solid internal infrastructure.

Furthermore, your internal systems have to be secure. Failing to properly secure your system only increases the likeliness of someone breaking into it, stealing your data, or even taking control of your autonomous tractors.

Technology use Cases for Smart Farming

Edge computing has already begun to make its mark on a variety of different industries across the planet. From manufacturing and distribution to healthcare and retail services, there seems to be few places left in which the presence of some form of edge computing cannot be felt. One of the areas in which edge computing is beginning to transform the landscape is farming and agriculture.

As we move into an age in which technology is becoming increasingly interconnected with both our infrastructure and our work and personal lives, there are various reasons as to why the adoption of new, cutting edge innovations such as edge computing, automation and artificial intelligence technologies. When you consider some of the challenges we face today, and those we will face in the not so distant future, utilizing technologies such as these in an industry as essential to us all as farming and agriculture makes perfect sense.

Video Analytics

Smart farming all about collecting the right data and using it to optimize your resource-planning and operation. Artificial Vision technologies using artificial intelligence are making waves in the agriculture sector just like they are in any other. Smart Agriculture is a major use case for vision-based automation and data analytics applications, driven by drones and vision based harvesting, weeding and so on. Deploying compute capabilities using ventless industrial PC's at the edge allows for on-site analytics and quick access to graphic-heavy data and analytics.

Besides edge computing devices, deployment of 4G/LTE/5G connectivity is another key factor in evolution of high-resolution visual data collection in remote environment. Industrial Communication Gateways with integrated LTE connectivity offer a great on-site compute platform along with cellular communication with cloud.

Environmental Monitoring

One of the biggest advantages that edge computing has brought to farming and agriculture over the past few years is the ability to remotely monitor different aspects of a farm's agricultural operations. Networks of sensors, ranging from several sporadically placed sensors to thousands of connected devices monitoring aspects such as soil, weather and humidity and temperature conditions as well as acidity and pH levels.

Edge computing allows for the data necessary in delivering such solutions to be generated and collected much closer to the source while also enabling some data processing operations to be performed in the edge devices themselves. Allowing farmers and agricultural workers detailed insights about their operational environments is a big selling point with any new technology, and edge computing is no different.

Robotics

While neither robotics technologies or artificial intelligence systems have quite given us the android companions science fiction has predicted for so long, we are now getting to a point in time where such robotic iterations don't seem so much like science fiction anymore and appear to us more as feasible possibilities.

While robotics has been an active part of many industries for decades now, car manufacturing, for example, it has also been slow to gain traction elsewhere. With the help of edge computing and IoT device networks, this is no longer the case.

Within agriculture, the recent emergence of smart farming frameworks has promoted the use of edge computing and Internet of Things (IoT) systems devices, and laid the groundwork to enable IoT-driven robotics to be developed for smart farms. These systems are capable of being programmed to perform automated tasks such as picking vegetables as well as spraying plants and crops, with more applications currently being developed.

Automation

The use of virtualization, automation, and Big Data have been the key drivers for what has become known as the fourth industrial revolution, and so it is no surprise that automation is playing a growing role in many smart farming and agricultural enterprises. automated robotics, as well as soil injections, heaters, and lighting can all be utilised through the use of edge computing and automation systems.

When coupled with the various remote monitoring applications and software now available to smart farms and agricultural enterprises, automation can become a powerful tool to farmers. The many benefits of automation continue to grow as do its applications within smart farming and, with systems such as 5G wireless networks and powerful machine learning algorithms also currently being developed, it seems automation's role in smart farming and agriculture is just beginning.

Vertical Farming

We as a species live on a planet of finite resources, and one of those finite resources is farmable land. Soil degradation is a growing problem around the world and the amount of farmable land we have is, unfortunately, slowly decreasing. In response to this growing crisis, scientists and farmers from around the world have developed what has become known as vertical farming. Vertical farming involves using the data collected from a network of IoT sensors and devices to optimize the growing of food and plants without the need for farm land.

In vertical farms for example, moisture levels are controlled with a network of sensors that constantly monitor a mist that surrounds the plant. Using edge computing, much of the data processing involved in such operations can be done on the edge devices themselves, without the need to be sent to the cloud, further adding to the benefits of such systems within farming environments.

While we continue to move into the Fourth Industrial Revolution, analytics technologies, automation, 5G, and AI and machine learning technologies will all begin to become more prominent and integrated into smart farms and agriculture.

Development of Automated Devices for the Monitoring of Insect Pests

Monitoring of adult insects with traps is considered a standard activity in integrated pest management (IPM) and early warning detection, helping to optimize control or eradication operations through observing the presence and/or variation of a pest population in the field. The data collected have been used to provide knowledge or warning to farmers and other agricultural stakeholders and allowed to incorporate spatial and temporal variability of the pest populations sampled in the field. This was key to accurately respond to the observed variations with precision treatments.

The increasingly widespread use of information and communication technologies (ICT) such as sensor networks, communication devices, internet things, and big data management and simulation software has opened many opportunities to modern agricultural production systems. In particular, automated or electronic traps (briefly e-traps or smart traps) have been suggested as part of ICT tools for pest monitoring.

Concept and Design of Automated Devices

The catching module of the e-trap is adapted from existing or newly developed devices and can include an attractant that increases attractiveness or makes the trap more selective. It contains the sensor, which allows detecting and counting of the target insects. Most used sensors are optical but different types, for example activated by acoustic vibrations or electric fields, were also used.

Photo-interruption sensors are activated by the modification or interruption of incident light (visible or infrared) due to falling objects such as moving insects and aim at counting the number of times a target insect entered in the trap. It was among the first systems to be used in combination with electronic components, transmitting the counts to a computer. In this case, the counted insects are not identified; therefore, the device must be specific to avoid erroneous counts caused by non-target specimens entering the trap.

An image sensor detects and conveys light waves to produce an image. The camera must have a minimum resolution enabling correctly classification of the captured insect. It has been calculated the minimum resolution to be 2 megapixels to capture a recognizable image. The number of photos is limited by the system's power consumption although, for the intended use here, one or a few photographs per day are easily supported.

E-traps equipped with this sensor type can work at different levels of automation. In semiautomatic systems, the sensor collects the images at certain times and transmits them via the internet to a remote operator who can check and count insects directly watching the image from a remote device and this can be done in real time. In case of a fully automated system, target individuals in the image are recognized and counted by image classifier algorithms, mostly based on machine learning or deep learning techniques. Various discriminating features, for example, the dimension or proportion of the body, can be used to differentiate the target species from other species.

Other optoacoustic sensors analyse the flow of light modulated by the wingbeats of an insect entering the trap. In this case, the wingbeat represents a sort of biometric signature that allows us to discriminate the target from non-target specimens.

Various non-optical sensors were tested for insect trapping, able to recognize the bioacoustics vibrations or electric field modifications resulting from locomotion and feeding behaviours of the target pests. Additional sensors to measure temperature, humidity, rainfall etc. can be included in the e-traps.

A microcomputer represents the core of the e-trap electronic system, collecting and recording the data from sensors and sending them to a network module. A unit for the remote connection is an important component of the automated trapping device to transmit data to the cloud usually using a 3G-4G-5G connection, secured by internet service providers. When many e-traps are deployed, costs for internet access can increase. In this case, e-traps can be connected to each other as nodes in wireless networks with various configurations and a central node transmitting data to the cloud. Different communication protocols have been used so far depending on the local conditions in the deployment of traps. For example, ZigBee had a long distance range but a low data transmission rate, whereas, Wi-Fi/WiMax had a higher data transmission rate but was more energy demanding. A source of electricity is necessary to sustain all electronic components; when the power grid is not directly available, batteries and solar panels can supply power.

A graphical user interface (GUI) allows to read and interpret the information remotely. Additional software such as decision support systems (DSS), can be fed with data coming from automated traps and provide information on the selection of control tools or assist the operator in a spraying process.

Automated Devices: Practical Applications

The first automated devices, equipped with bioacoustic sensors, image sensors or with light emitting diodes and the related receiving sensor have been designed for automatic trapping of Coleoptera. Earlier sensors were applied to passive grain probe traps to provide continuous monitoring of stored-grain beetles within large volumes of stored products. A modified automatic pheromone trap for the boll weevil *Anthonomus grandis* Boheman was tested for the season monitoring of the weevil in cotton fields. Various traps baited with the specific aggregation pheromone were proposed for the monitoring of the red palm weevil, *Rhynchophorus ferrugineus* (Olivier) in urban environments or bark beetles in forests. A different approach was used for the early detection of alien wood-boring beetles. In this case, a multi-funnel trap baited with a blend of generic lures attractive to many wood-boring species was equipped with a camera; the images were inspected remotely for insect identification with a different level of accuracy depending on family and genus of the trapped beetles.

Another example of the automatic image-based trap was developed for small-bodied insects mainly aimed for greenhouses. An automatic pheromone trap for counting the bean bug *Riptortus clavatus* (Thunberg) utilized a different detecting system consisting of two rollers placed at the same distance of the size of the insect, which upon entering touches both of them and generates an electric arc by striking the insect. The counts are transmitted by using a mobile phone connected to the trap.

Among Lepidoptera, various automated trapping devices baited with sex pheromones were established to monitor *Cydia pomonella* (L.) in apple orchards equipped with a different type of sensors. An e-trap based on an optical sensor was developed based on a modified commercial model by allocating at the top of the trap a mobile phone with a camera taking images of the sticky surface placed at the bottom side and sending it immediately to the remote server for a visual evaluation. Other developed traps used a visible or infrared light emitting diodes equipped with optoelectronic sensors able to count the trapped moths by detecting the light interruption caused by individuals falling through the funnel of a bucket trap. Another device called z-trap was equipped with a metallic coil, which was able to identify the species of insect flying into the trap based on the amount of electric current discharged when an insect touches the coil. A monitoring automated device equipped with infrared sensors was proposed for counting *Spodoptera litura* (F.) moths entering in pheromone trapping tubes.

Many types of research focused on the development of automated monitoring tools for fruit flies. An electronic device for *Bactrocera dorsalis* (Hendel) employed an infrared interruption sensor to count attracted flies entering the trap baited with the attractant methyl eugenol through an electronic funnel. A wireless automatic trap was developed by modifying a McPhail model baited with the specific pheromone attractants for the monitoring of *Ceratitis capitata*Wiedemann and *Bactrocera oleae* (Gmelin), equipped with a camera to capture images of insects; here a software automatically identifies and counts fruit fly entering into the trap along a transparent funnel.

Other technologies for e-monitoring used a location-aware system based on a real-time wireless multimedia sensor network (WMSN). The system, through a semi-automatic trapping and insect counting, is able to acquire and transmit data to a remote server feeding a DSS that performed the final optimization of the control treatments. In particular, two of these e-trap models were based on the wireless transmission of images of trapped fruit flies on the glue surfaces of the e-traps, checked remotely by an entomologist. Their development and validation focused on following specific species in different agro-ecosystems: Medfly in peach orchards, cherry fruit fly in cherry orchards, olive fruit fly in olive orchards and the Ethiopian fruit fly in melons growing in plastic tunnels, modifying delta-type traps or yellow sticky panels, baited with different pheromone or food attractants, depending on the target fruit fly. The verification of the reliability of the data obtained was performed comparing the captures counted on the transmitted image with the captures checked by a human in the e-trap in the field. The captures of the flies checked remotely usually showed a similar numerical trend and the number of flies caught in the e-traps was similar to the number obtained with the standard manually-checked traps.

Automatic devices based on infrared sensors were also developed for automatic trapping of *C. capitata* to optimize control applications frequency. A McPhail trap based on optoelectronic sensors detecting differences in the optoacoustic spectrum of insect wingbeat resulting from entering insects into the trap was developed for *B. oleae*.

Current Constrains and Future Perspectives of Automated Devices for Insect Trapping

Traps, which frequently include attractants, are very powerful tools to attract and capture specimens of the target pest. Sometimes, these traps are species-specific but usually require trained personnel to discriminate the target from similar-looking non-target species that can be captured indeed. In many standard IPM programs, especially in perennial crops, different trap types are used for the monitoring and control of specific insect species, deployed as grids in the fields, checked and serviced manually through periodic visits of human operators.

Standard monitoring procedures involve the manual counting of trapped insects. The field survey of traps is done usually once a week, sometimes twice, for most insects. Each time, there are delays in data acquisition and analysis because inspectors must reach the trap in the field, count the insects from each device, enter data on paper then go back to the office, enter data in a spreadsheet and process them to send out to final users. This process has been in part shortened, inserting data directly in the field with a portable device that can communicate remotely with a server that stores the data automatically. However, assessing traps results in a delay that affects the time necessary to make a decision.

The biosecurity surveillance activities of alien insects also include extensive use of baited traps both at points-of-entry for imported commodities to capture insects before they become established and in the context of post-border surveillance and containment. In these cases, traps must be monitored frequently as specific action must be taken immediately when a quarantine species is detected and delays can increase the chances for the establishment of a new invasive species. Here, automated traps together with other technologies can play a major role in reducing the response times as explained above.

By themselves, manual traps are relatively simple tools to be managed and have a very low cost. However, intensive labour and transportations for installation, maintenance and periodic check of traps by skilled personnel represent most of the total monitoring costs. The service of large monitoring networks can become very expensive, especially in areas where access to the traps is difficult. For this reason, often the number of monitoring traps positioned in the field is greatly reduced with a consequent loss of information on the spatiotemporal dynamics of pests and inaccurate control applications or missed alerts.

What are the potential advantages of automatic trapping systems? Data available on real-time are easily represented in time and space and can automatically feed a DSS for the optimization of control methods. There is a reduction in human and transportation costs because traps are no longer checked every week but less often. The monitoring in remote or inaccessible areas is facilitated. When positioned the trap can remain for a long time without an operator going into the field to check it and can transmit data at desired time intervals. There is increased efficiency in area-wide programs, where many traps are located in a large territory: Here e-traps represent an efficient method to get all needed information in real time. Coupled to a DSS, the automatic trapping may improve the applications of precision agriculture to pilot the operators in doing control interventions at the right time and place.

Real-time monitoring allows being very efficient in some particular situations like an early warning of invasive pests. E-traps producing images have the advantage that the operator can check frequently if suspected alien insects are in the traps and if a secure identification cannot be done remotely, go timely in the field for confirmation. In this case, the advantage of the e-trap is related to the possibility to give a remote secure identification. E-traps baited with a specific pheromone coupled with image analysis software, can be highly efficient in detecting correctly the target species. The same applies for other highly specific sensors such as those using wing beat frequencies; on the contrary, low specific sensors giving simply a ping for something arriving in the trap are not useful, unless the trap is highly specific.

From an electronic and informatics point of view, smart traps are mature and reliable technology. Over the last few years, the electronic parts have reached increasingly higher performances and lower costs, allowing the miniaturization of components. Therefore, the e-trap itself is not very expensive and labour costs are strongly reduced. Other benefits in terms of cost savings from reduced insecticide use and insect damages should be specifically assessed. However, possible economic constraints are costs for the development of a DSS or image analysis software. Furthermore, technologies should be tested in field conditions to prevent problems like insufficient battery capacity to make the device work properly, failure in data transmission to the cloud and weather conditions (rains or extreme temperatures) that can affect the e-trap field operation over time.

Modern agriculture is facing a huge technological transformation. Drones, remote sensing, intelligent decision support systems, internet of things introduced us in the smart agriculture concept and methodology in which we can place the automatic traps.The development of automated monitoring devices requires specific skills and a multidisciplinary team with computer engineers, ICT specialists, entomologists and IPM professionals. For the dissemination of this type of technological applications the end users, whether they are agricultural companies, producer associations, private consultants or public bodies, must be willing to invest in maintaining the automatic monitoring networks including the costs for internet access and the data retention online. There is a big potential for the utilization of such an innovative tool, especially in high-value crops or when high labour costs. Examples of fully automated traps mostly for a moth or fruit fly pests using different sensor typology are currently available in the market. As for other ICTs, the perspective for smart traps to be widely used in the farming practice in the near future must be considered substantial.

Applications of GPS in Agriculture

Global Positioning System (GPS) has forever revolutionized location, monitoring, navigation and other related complex applications. Today, GPS technology has been transferred partly in civil sphere, finding use in many fields such as auto transportation, rail, ship and aircraft, construction equipment, equipment monitoring and surveillance, agriculture, cartography, medicine and so on.

GPS is a network of satellites orbiting around the Earth, they form the GPS network. The satellites were placed into the space to be used for military purposes by the Department of Defense U.S. Army, but today it can be used by civilians. Today, the satellites are the responsibility of the United States Air Force and the maintenance and replacement exceeds the amount of U.S. $ 700 million annually. U.S. government does not charge for services using GPS, the only costs incurred by users as those relating to purchase GPS receivers.

Although GPS technology has seen, including our country, a rapid and wide utilization, we can see that the general principles of this technology are relatively unknown and sometime ignored. This situation could lead to overstated performances of GPS technology ("if this was measured with the GPS it means that the results cannot be wrong") or to an inefficient and incomplete usage of possibilities.

GPS is a worldwide radio navigation system formed from a constellation of 24 satellites placed in orbit around Earth and associated ground stations. Moreover, these satellites are evenly distributed as to ensure a uniform coverage of the entire globe. Such a distribution suppose the existence of 6 orbits with 4 satellites. The 24 satellites with the trajectories are composing the space segment of the GPS. Technical characteristics of the GPS satellites refers to altitude (20200 km), time (12 hours), frequency 1557 MHz, 1228 MHz, data navigation (4D, X, Y, Z, t) and so on.

The control segment is one of the most important parts of the whole system. It controls the satellites, checks the satellites time, calculates the orbits and the time corrections and navigation data and the content of each message received from the satellites.

Segment consists of a main station (Master Control Station / MCS) located in Colorado Springs, three monitoring stations and ground antennas in Kwajalein, Ascension Island and Diego Garcia and two monitoring stations in Colorado Springs and Hawaii.

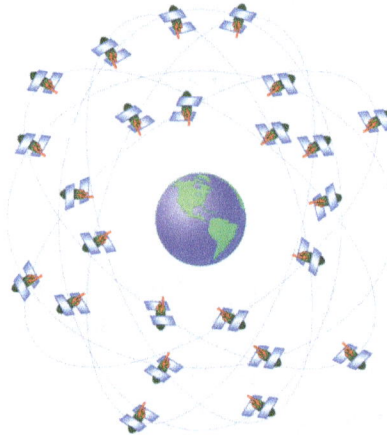

GPS satellite constellation.

Ground Stations.

The monitoring stations are receiving signals from satellites and send them to the master station along with meteorological data. Here, the received data is processed resulting navigation data and time corrections. Data is then redirected to antenna stations being transmitted under a message format to each satellite.

To calculate the exact position the system uses the triangulation principle, being necessary to know the coordinates from three satellites.

GPS uses this artificial stars to calculate the terrestrial position of objects with a precision measured in meters. In fact, with advanced forms of GPS, DPS (Differential Global Positioning System) we can do measurements with an accuracy of a few centimeters.

Materials and Methods

One of the areas where the GPS communication technology can be used successfully, is surprisingly the agriculture, especially that this is a very traditional industry. The integration of our country in European Union imposes the following of European standards and ranging to the Common Agricultural Policy, which is difficult to accomplish without the use of new technologies. To benefit from the European funds, Romania must have high accuracy agricultural cadastral plans; accurate digital agricultural plans and maps; detailed databases for cadastral agricultural data; precise measurements of agricultural and nonagricultural areas; permanent means of measuring the areas cultivated with different types of cultures and crops; permanent control of areas with different types of crops; means of updating permanently the agricultural infrastructure through the introduction of the newest technology, hence answering the stringent requests imposed by EU. Using a computer system based on GPS and DGPS technologies will allow to achieve quickly and efficiently the requirements imposed by the EU.

The Integrated System for Administration and Control (IACS) was set-up in Romania, to monitor some types of financial aids. For this is necessary to collect data regarding terrain nature and surface, crop area and harvests. The usage of GPS/DGPS allows for an efficient way to obtain accurate data for administrating and control of financial aids. The GPS/DGPS applied in agriculture represents the main tool for real time management of specific agricultural information and to help in allocate correctly funds, both locally and nationally.

A correct assessment of GPS position measurements in the field measurement techniques must be based first on an analysis of both-related opportunities regarding extreme lengths domain, on the other hand the accuracies obtained with these techniques and no ultimately the criteria for equipment and personnel costs. This comparison can be traced in figure.

Classical methods for determining the length does not exceed 60 km with precision of about 0.25 to 0.30 m. Around the same field length are inertial methods that fall transit methods.

At the other extreme of distances it is the interferometry with very long basis with laser range measurements from satellites. Both methods are extremely sophisticated and expensive, difficult to use in daily activities, especially in agriculture.

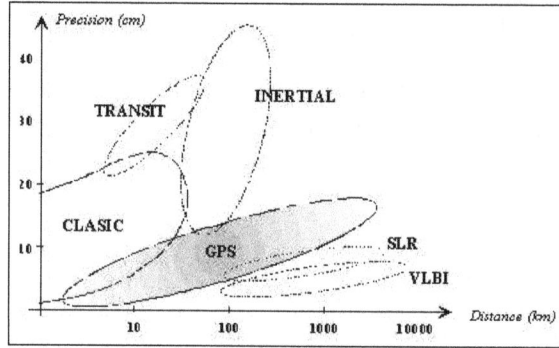

Comparison of measurement techniques.

If we refer to accuracies determinations, we note that classical field and the GPS distances are comparable only in less than 15 km, because over this value, the GPS technique is more accurate. Another detail to be considered is that the of the ease of handling. As important steps were made in miniaturization receivers, they practically become extremely easy to use, regardless of the position on the globe or relief conditions. In our country,the use in agriculture of this technology is at the beginning, being used in most cases for mapping land parcels.

For example, in developed countries for mapping land according to harvest achieved a software is used, to obtain express crop yield maps obtained from small units of area.

In the developed countries the mapping of land according to the harvest is achieved through a specialized software. This will greatly help in forecasting the possible productive potential of the parcels.For example using such technology the following map can be obtained.

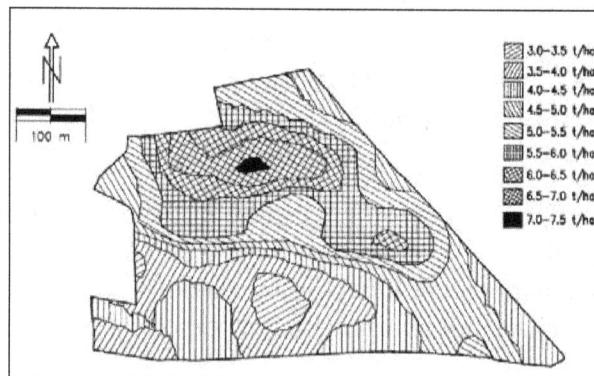

Productive potential of a parcel map.

This is a map of productive potential of a parcel is of 16.6 hectares, sown with wheat and with an average production of 5.2 t/ha . This example is for a farm in Germany.

References

- Iot-applications-in-agriculture: iotforall.com, Retrieved 16 June, 2019

- Iot-applications-in-agriculture-the-potential-of-smart-farming-on-the-current-stage, datadriveninvestor: medium.com, Retrieved 26 January, 2019

- Iot-in-agriculture-five-technology-uses-for-smart: dzone.com, Retrieved 25 August, 2019

- 5-edge-computing-use-cases-smart-farming-agriculture: lanner-america.com, Retrieved 16 April, 2019

- Development-of-automated-devices-for-the-monitoring-of-insect-pests: agriculturejournal.org, Retrieved 18 July, 2019

Permissions

We would like to thank the editorial team for lending their expertise to make the book truly unique. They have played a crucial role in the development of this book. Without their invaluable contributions this book wouldn't have been possible. They have made vital efforts to compile up to date information on the varied aspects of this subject to make this book a valuable addition to the collection of many professionals and students.

This book was conceptualized with the vision of imparting up-to-date and integrated information in this field. To ensure the same, a matchless editorial board was set up. Every individual on the board went through rigorous rounds of assessment to prove their worth. After which they invested a large part of their time researching and compiling the most relevant data for our readers.

The editorial board has been involved in producing this book since its inception. They have spent rigorous hours researching and exploring the diverse topics which have resulted in the successful publishing of this book. They have passed on their knowledge of decades through this book. To expedite this challenging task, the publisher supported the team at every step. A small team of assistant editors was also appointed to further simplify the editing procedure and attain best results for the readers.

Apart from the editorial board, the designing team has also invested a significant amount of their time in understanding the subject and creating the most relevant covers. They scrutinized every image to scout for the most suitable representation of the subject and create an appropriate cover for the book.

The publishing team has been an ardent support to the editorial, designing and production team. Their endless efforts to recruit the best for this project, has resulted in the accomplishment of this book. They are a veteran in the field of academics and their pool of knowledge is as vast as their experience in printing. Their expertise and guidance has proved useful at every step. Their uncompromising quality standards have made this book an exceptional effort. Their encouragement from time to time has been an inspiration for everyone.

The publisher and the editorial board hope that this book will prove to be a valuable piece of knowledge for students, practitioners and scholars across the globe.

Index

www.ingramcontent.com/pod-product-compliance
Lightning Source LLC
Chambersburg PA
CBHW061955190326

41458CB00009B/2878